PENGUIN BOOKS

MENAGERIE MANOR

Gerald Durrell was born in Jamshedpur, India, in
1925. In 1928 his family returned to England and in
1933 they went to live on the Continent. Eventually
they settled on the island of Corfu, where they lived
until 1939. During this time he made a special study
of zoology, and kept a large number of the local
wild animals as pets. In 1945 he joined the staff of
Whipsnade Park as a student keeper. In 1947 he
financed, organized and led his first animal-
collecting expedition to the Cameroons. He has also
made expeditions to Guyana, Paraguay, Sierra
Leone and Argentina. In 1963 he and his wife went
to New Zealand, Australia and Malaya to film a TV
series, *Two in the Bush*, in conjunction with the B.B.C.
Natural History Film Unit. In 1958 he founded the
Jersey Zoological Park, of which he is the director,
and in 1964 he founded the Jersey Wildlife Preservation Trust. Gerald Durrell's other books include *A
Zoo in my Luggage*, *The Whispering Land*, *Three Singles
to Adventure*, *The Drunken Forest*, *The Bafut Beagles*,
My Family and Other Animals and *Encounters with
Animals* (all published in Penguins). His latest
books are *Catch Me a Colubus* (1972), *Beasts in My
Belfry* (1973), *The Talking Parcel* (1974) and *The
Stationary Ark* (1976).

GERALD DURRELL

MENAGERIE MANOR

WITH ILLUSTRATIONS BY

Ralph Thompson

PENGUIN BOOKS

Penguin Books Ltd, Harmondsworth, Middlesex, England
Penguin Books, 625 Madison Avenue, New York, New York 10022, U.S.A.
Penguin Books Australia Ltd, Ringwood, Victoria, Australia
Penguin Books Canada Ltd, 2801 John Street, Markham, Ontario, Canada L3R 1B4
Penguin Books (N.Z.) Ltd, 182–190 Wairau Road, Auckland 10, New Zealand

—

First published in Great Britain by Rupert Hart-Davis 1964
First published in the United States by The Viking Press, Inc., New York, 1964
Published in Penguin Books 1967
Reprinted 1967, 1968, 1971, 1972, 1973, 1974, 1975, 1976, 1977, 1979
This edition published by Penguin Books Inc.
by arrangement with The Viking Press, Inc.

—

—

Made and printed in Great Britain
by Richard Clay (The Chaucer Press) Ltd,
Bungay, Suffolk
Set in Monotype Garamond

CONTENTS

EXPLANATION

Dear Sir,
We should like to draw your attention to the fact that your account
with us is now overdrawn . . .

MOST children at the tender age of six or so are generally full of the most impractical schemes for becoming policemen, firemen or engine drivers when they grow up, but when I was that age I could not be bothered with such mundane ambitions. I knew exactly what I was going to do: I was going to have my own zoo. At the time this did not seem to me (and still does not seem) a very unreasonable or outrageous ambition. My friends and relatives – who had long thought that I was mental owing to the fact that I evinced no interest in anything that did not have fur, feathers, scales or chitin – accepted this as just another manifestation of my weak state of mind. They felt that, if they ignored my oft-repeated remarks about owning my own zoo, I would eventually grow out of it. As the years passed, however, to the consternation of my friends and relatives, my resolve to have my own zoo grew greater and greater, and eventually, after going on a number of expeditions to bring back animals for other zoos, I felt the time was ripe to acquire my own.

From my last trip to West Africa I had brought back a considerable collection of animals which were ensconced in my sister's suburban garden in Bournemouth. They were there, I assured her, only temporarily because I was completely convinced that any intelligent council, having a ready-made zoo planted on its doorstep, would do everything in its power to help one by providing a place to keep it. After eighteen months of struggle, I was not so sure of the go-ahead attitude of local councils, and my sister was utterly convinced that her back garden would go on for ever looking like a scene out of one of the more flamboyant Tarzan pictures. At last, bogged down by the constipated mentality of local government and frightened off by the apparently endless rules and regulations under which every free man in Great Britain has to suffer, I decided to investigate the possibility of starting my zoo in the Channel Islands. I was given an introduction to one Major Fraser who, I was assured, was a broad-minded, kindly soul, and would show me round the Island of Jersey and point out suitable sites.

My wife, Jacquie, and I flew to Jersey where we were met by Hugh Fraser who drove us to his family home, probably one of the most beautiful manor houses on the Island: here was a huge walled garden dreaming in the thin sunlight; a great granite wall, thickly planted with waterfalls of rock plants; fifteenth-century arches, tidy lawns and flower-beds brimming over with colour. All the walls, buildings and outhouses were of beautiful Jersey granite which contains all the subtle colourings of a heap of autumn leaves and they glowed in the sunshine and seduced me into making what was probably the silliest remark of the century. Turning to Jacquie, I said:

'What a marvellous place for a zoo.'

If Hugh Fraser, as my host, had promptly fainted on the spot, I could scarcely have blamed him: in those lovely surroundings the thought of implanting the average person's idea of a zoo (masses of grey cement and steel bars) was almost high treason. To my astonishment Hugh Fraser did not faint but merely cocked an inquiring eyebrow at me and asked whether I really meant what I said. Slightly embarrassed, I replied that I had meant it, but added hastily that I realized it was impossible. Hugh said he did not think it was as impossible as all that. He went on to explain that the house and grounds were too big for him to keep up as a private individual, and so he wanted to move into a smaller place in England. Would I care to consider renting the property for the purpose of establishing my zoo? I could not conceive a more attractive setting for my purpose, and, by the time lunch was over, the bargain had been sealed and I was the new 'Lord' of the Manor of Les Augres in the Parish of Trinity.

The alarm and despondency displayed by all who knew me when I announced this, can be imagined. The only one who seemed relieved by the news was my sister, who pointed out that, although she thought the whole thing was a hare-brained scheme, at least it would rid her back garden of some two hundred assorted denizens of the jungle, which were at that time putting a great strain on her relationship with the neighbours.

To complicate things even more, I did not want a simple straightforward zoo, with the ordinary run of animals: the idea behind my zoo was to aid in the preservation of animal life. All over the world various species are being exterminated or cut down to remnants of their former numbers by the spread of civilization. Many of the larger species are of commercial or touristic value, and, as such, are receiving the most attention. Yet, scattered about all over the world are a host of fascinating small mammals, birds and reptiles, and scant attention is being paid to their preservation, as they are neither edible nor wearable, and of little interest to the tourist who demands lions and rhinos. A great number of these are island fauna, and as such their habitat is small. The slightest interference with this, and they will vanish for ever: the casual introduction of rats, say, or pigs could destroy one of these island species within a year. One has only to remember the sad fate of the dodo to realize this.

The obvious answer to this whole problem is to see that the creature is adequately protected in the wild state so that it does not become extinct, but this is often easier said than done. However, while pressing for this protection, there is another precaution that can be taken, and that is to build up under controlled conditions breeding stocks of these creatures in parks or zoos, so that, should the worst happen and the species become extinct in the wild state, you have, at least, not lost it for ever. Moreover, you have a breeding stock from which you can glean the surplus animals and reintroduce them into their original homes at some future date. This, it has always seemed to me, should be the main function of any zoo, but it is only recently that the majority of zoos have woken up to this fact and tried to do anything about it. I wanted this to be the main function of my zoo. However, like all altruistic ideas, it was going to cost money. It was, therefore, obvious the zoo would have to be run on purely commercial lines to begin with, until it was self-supporting. Then one could start on the real work of the zoo: building up breeding stocks of rare creatures.

So this is the story of our trials and tribulations in taking the first step towards a goal which I think is of great importance.

CHAPTER ONE

MENAGERIE MANOR

Dear Mr Durrell,
I am eighteen years old strong in wind and limb having read your
books can I have a job in your zoo . . .

IT is one thing to visit a zoo as an ordinary member of the public but quite another to own a zoo and live in the middle of it: this at times can be a mixed blessing. It certainly enables you to rush out at any hour of the day or night to observe your charges, but it also means that you are on duty twenty-four hours a day, and you find that a cosy little dinner party disintegrates because some animal has broken its leg, or because the heaters in the Reptile House have failed, or for any one of a dozen reasons. Winter, of course, is your slack period, and sometimes days on end pass without a single visitor in the grounds and you begin to feel that the zoo is really your own private one. The pleasantness of this sensation is more than slightly marred by the alarm with which you view the mounting of your bills and compare them to the lack of gate-money. But in the season the days are so full and the visitors so numerous that you hardly seem to notice the passing of time, and you forget your overdraft.

The average zoo day begins just before dawn; the sky will be almost imperceptibly tinged with yellow when you are awakened by the bird-song. At first, still half asleep, you wonder whether you are in Jersey or back in the tropics, for you can hear a robin chanting up the sun, and, accompanying it, the rich, fruity, slightly hoarse cries of the touracous. Then a blackbird flutes joyfully, and as the last of his song dies the white-headed jay thrush bursts into an excited, liquid babble.

As the sky lightens, this confused and cosmopolitan orchestra gathers momentum, a thrush vies with the loud, imperious shouts of the seriamas, and the witches' cackle from the covey of magpies contrasts with the honking of geese and the delicate, plaintive notes of the diamond doves. Even if you survive this musical onslaught and can drift into a doze again, you are suddenly and rudely awakened by something that resembles the strange, deep vibrating noise that a telegraph pole makes in a high wind. This acts upon you with the same disruptive effect of an alarm clock, for it is the warning that Trumpy has appeared, and if you have been foolish enough to leave your window wide open you have to take immediate defensive action. Trumpy is a grey-winged trumpeter, known to his more intimate ornithologist friends as *Psophia crepitans*. His function in the Zoo is three-fold – combined guide, settler-in and village idiot. He looks, to be frank, like a badly made chicken, clad in sombre plumage as depressing as Victorian mourning: dark feathers over most of his body and

what appears to be a shot-silk cravat at his throat. The whole ensemble is enlivened by a pair of ash-grey wings. He has dark, liquid eyes and a high, domed forehead which argues a brain-power which he does not possess.

Trumpy – for some reason best known to himself – is firmly convinced that his first duty of each day should be to fly into one's bedroom and acquaint one with what has been going on in the Zoo during the night. His motives are not entirely altruistic, for he also hopes to have his head scratched. If you are too deeply asleep – or too lazy – to leap out of bed at his greeting cry, he hops from the window-sill on to the dressing-table, decorates it extravagantly, wags his tail vigorously in approval of his action, and then hops on to the bed and proceeds to walk up and down, thrumming like a distraught cello until he is assured that he has your full attention. Before he can produce any more interesting designs on the furniture or carpet, you are forced to crawl out of bed, stalk

and catch him (a task fraught with difficulty, since he is so agile and you are so somnambulistic), and push him out on to the window-ledge and close the window so that he cannot force his way in again. Trumpy now having awakened you, you wonder sleepily whether it is worth going back to bed, or whether you should get up. Then from beneath the window will come a series of five or six shrill cries for help, apparently delivered by a very inferior soprano in the process of having her throat cut. Looking out into the courtyard, on the velvet-green lawns by the lavender hedge, you can see an earnest group of peahens searching the dewy grass, while around them their husband pirouettes, his shining and burnished tail raised like a fantastic, quivering fountain in the sunlight. Presently he will lower his tail, and, throwing back his head, will deafen the morning with his nerve-shattering cries. At eight o'clock the

Gorilla

staff arrive, and you hear them shouting greetings to each other, amid the clank of buckets and the swish of brushes, which all but drowns the bird-song. You slip on your clothes and go out into the cool, fresh morning to see if all is right with the Zoo.

In the long, two-storied granite house – once a large cider press and now converted for monkeys and other mammals – everything is bustle and activity. The gorillas have just been let out of their cage, while it is being cleaned, and they gallop about the floor with the exuberance of children just out of school, endeavouring to pull down the notices, wrench the electric heaters from their sockets or break the fluorescent lights. Stefan, brush in hand, stands guard over the apes, watching with a stern eye, to prevent them from doing more damage than is absolutely necessary. Inside the gorillas' cage Mike, rotund and perpetually smiling, and Jeremy, with his Duke of Wellington nose and his barley-sugar-coloured hair, are busy, sweeping up the mess that the gorillas' tenancy of the previous day entailed and scattering fresh, white sawdust in snowdrifts over the floor. Everything, they assure you, is all right: nothing has developed any malignant symptoms during the night. All the animals, excited and eager at the start of a new day, bustle about their cages and shout 'Good morning' to you. Etam, the black Celebes ape, looking like a satanic imp, clings to the wire, baring his teeth at you in greeting and making shrill, chuckling noises. The woolly-coated orange-eyed mongoose lemurs bound from branch to branch, wagging their long thick tails like dogs, and calling to each other in a series of loud and astonishingly pig-like grunts. Farther down, sitting on his hind legs, his prehensile tail wrapped round a branch, and surveying his quarters with the air of someone who has just received the freedom of the city, is Binty, the binturong, who suggests a badly made hearthrug, to one end of which has been attached a curiously

oriental-like head with long ear-tufts and circular, protuberant and somewhat vacant eyes. The next-door cage appears to be empty, but run your fingers along the wire and a troupe of diminutive marmosets comes tumbling out of their box of straw, twittering and trilling like a group of canaries. The largest of these is Whiskers, the emperor tamarin, whose sweeping snow-white Colonel Blimp moustache quivers majestically as he gives you greeting by opening wide his mouth and vibrating his tongue rapidly up and down.

Upstairs, the parrots and parakeets salute you with a caco-phony of sound: harsh screams, squeakings resembling un-oiled hinges, and cries that vary from 'I'm a very fine bird', from Suku, the grey parrot, to the most personal 'Hijo de puta', squawked by Blanco, the Tucuman Amazon. Farther along, the genets, beautifully blotched in dark chocolate on their golden pelts, move as quicksilver through the branches of their cage. They are so long and lithe and sensuous that they seem more like snakes than mammals. Next door, Queenie, the tree ocelot, her paws demurely folded, gazes at you with great amber eyes, gently twitching the end of her tail. A host of quick-footed, bright-eyed, inquisitive-faced mongooses patter busily about their cages, working up an appetite. The hairy armadillo lies supine on its back, paws and nose twitching, and its pink and wrinkled stomach heaving as it dreams sweet dreams of vast plates of food. You reflect, as you look at it, that it is about time it went on a diet again, otherwise it will have difficulty in walking, and you make a bet with yourself as to how many visitors that day would come to tell you that the armadillo was on its back and apparently dying: the record to date has been fifteen visitors in one day.

Outside, the clank of a bucket, the burst of whistling, herald the approach of Shep, curly-haired and with a most disarming

Emperor tamarin

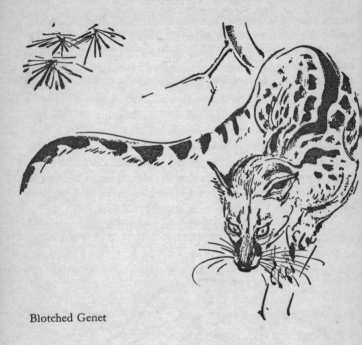

Blotched Genet

grin. As his real name is John Mallet it was inevitable that he should be called Shepton Mallet which, in turn, has degenerated into Shep. You walk up the broad main drive with him, past the long twelve-foot-high granite wall ablaze with flowering rock plants, and down to the sunken water-meadow where the swans and ducks swim eagerly to welcome him as he empties out the bucket of food at the edge of the water. Having ascertained from Shep that none of his bird charges have during the night sickened or died or laid eggs, you continue on your tour.

The Bird House is aburst with song and movement. Birds of every shape and colour squabble, eat, flutter and sing, so that the whole thing resembles a market or a fairground alight with bright colours. Here a toucan cocks a knowing eye at you and clatters his huge beak with a sound like a football rattle; here a black-faced lovebird, looking as though he had just come from a minstrel show, waddles across to his water-dish and proceeds to bathe himself with such vigour that all the other occupants of the cage receive the benefit of his bathwater; a pair of tiny, fragile diamond doves are dancing what appears to be a minuet together, turning round and round, bowing and changing places, calling in their soft, ringing voices some sort of endearments to each other.

You pass slowly down the house to the big cage at the end where the touracous now live. The male, Peety, I had hand-reared while in West Africa. He peers at you from one of the higher perches and then, if you call to him, he will fly down in a graceful swoop, land on the perch nearest to you and start to peck eagerly at your fingers. Then he will throw back his head, his throat swelling, and give his loud, husky cry: 'Caroo ... Caroo ... Caroo ... coo ... coo ... coo.' Touracous are really one of the most beautiful of birds. Peety's tail and wings are a deep metallic blue, while his breast, head and neck are a rich green, the feathering so fine and shining that it looks like spun glass. When he flies, you can see the undersides of his wings, which flash a glorious magenta red. This red is caused by a substance in the feathers, called turacin, and it is possible to wash it out of the feathers. Place a touracou's wing feather in a glass of plain water, and presently you will find the water tinged with pink, as though a few crystals of permanganate of potash had been dissolved in it. Having dutifully listened to Peety and his wife sing a duet together, you now make your way out of the Bird House.

Dodging the exuberant welcome of the chimpanzees, who prove their interest in your well-being by hurling bits of fruit – and other less desirable substances – with unerring accuracy through the wire of their cage, you walk to the Reptile House.

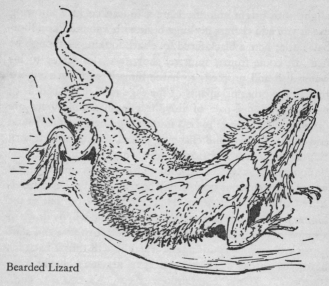

Bearded Lizard

Here in a pleasant temperature of eighty degrees the reptiles doze. Snakes regard you calmly with lidless eyes, frogs gulp as though just about to succumb to a bout of sobs, and lizards lie draped over rocks and tree-trunks, exquisitely languid and sure of themselves. In the cage which contains the Fernand skinks I had caught in the Cameroons, you can dig your hands into the damp, warm soil at the bottom of it and haul them out of their subterranean burrow, writhing and biting indignantly. They have recently shed their skins, and so they look as though they have been newly varnished. You admire their red, yellow and white markings on the glossy black background, and then let them slide through your fingers and watch as they burrow like bulldozers into the earth. John Hartley appears, tall and lanky, bearing two trays of chopped fruit and vegetables for the giant tortoises. The previous night had been a good feeding one, he tells you. The boa-constrictors had had two guinea pigs each, while the big reticulated python had engulfed a very large rabbit, and lies there bloated and lethargic to prove it. The horned toads, looking more like bizarre pottery figures than ever, had stuffed themselves on baby chickens, and, according to their size, the smaller snakes were busily digesting white rats or mice.

Round the back of the house are some more of the monkey collection that have just been let out into their outdoor cages: Frisky, the mandrill, massive and multi-coloured as a techni-coloured sunset, picks over a huge pile of fruit and vege-tables, grunting and gurgling to himself; farther along, Tarquin, the cherry-crowned mangabey, with his grey fur, his mahogany-coloured skull-cap and his white eyelids, goes carefully through the fur of his wife, while she lies on the floor of the cage as though dead. Periodically he finds a delectable fragment of salt in her fur and pops it into his mouth. One is reminded of the small boy who had witnessed

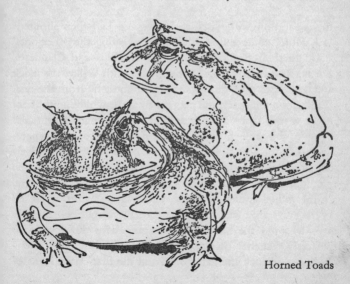

Horned Toads

this operation with fascinated eyes and had then shouted: 'Hi, Mum, come and see this monkey eating the other one.'

Up in their paddock the tapirs, Claudius and Claudette, portly, Roman-nosed and benign, play with Willie, the black and white cat, who guards the aviaries near by from the rats. Willie lies on his back and pats gently at the whiffling, rubbery noses of the tapirs as they sniff and nuzzle him. Eventually tiring of the game, he would rise and start to move off, where-upon one of the tapirs would reach forward and tenderly engulf Willie's tail in its mouth and pull him back, so that he would continue this game of which they never seemed to tire. In the walled garden the lions, butter-fat and angry-eyed, lie in the sun, while near them the cheetahs would be languidly asprawl amid the buttercups, merging with the flowers so perfectly that they became almost invisible.

At ten o'clock the gates open and the first coach-loads of

people arrive. As they come flooding into the grounds, everyone has to be on the alert, not, as you may think, to ensure that the animals do not hurt the people, but to ensure that the people do not hurt the animals. If an animal is asleep, they want to throw stones at it or prod it with sticks to make it move. We have found visitors endeavouring to give the chimpanzees lighted cigarettes and razor-blades; monkeys have been given lipsticks which, of course, they thought were some exotic fruit and devoured accordingly, only to develop acute colic. One pleasant individual (whom we did not catch, unfortunately) pushed a long cellophane packet full of aspirins into the chinchilla cage. For some obscure reason one chinchilla decided that this was the food it had been waiting for all its life and had eaten most of it before we came on the scene: it died the next day. The uncivilized behaviour of some human beings in a zoo has to be seen to be believed.

Now, there might be any one of fifty jobs to do. Perhaps you go to the workshop where Les, with his craggy face and bright eyes, is busy on some repair work or other. Les is one of those people who are God's gift to a zoo, for no job defeats him and his ingenuity is incredible. He is like a one-man building firm, for he can do anything from welding to dovetailing, from cementing to electrical maintenance. You discuss with him the new line of cages you are planning, their size and shape, and whether they should have swing doors, or whether sliding doors would be more convenient.

Having thrashed out this problem, you remember that one of the giant tortoises had to have an injection. On your way to deliver this, you pass an excited crowd of North Country

people round the mandrill cage, watching Frisky as he stalks up and down, grunting to himself, presenting now his vivid, savagely beautiful face, and now his multi-coloured rear to their eyes.

'Ee,' says one woman in wonderment, 'you can't tell front from back.'

Lunchtime comes and so far the day has progressed smoothly. As you sit down to eat, you wonder if there will be a crisis during the afternoon: will the ladies' lavatories overflow, or, worse still, will it start to rain and thus put off all the people who are intending to visit the Zoo? Lunch over, you see that the sky is, to your relief, still a sparkling blue. You decide to go down and look at the penguin pond, for which you have certain ideas of improvement.

You scuttle surreptitiously out of the house, but not surreptitiously enough, for both your wife and your secretary catch you in rapid succession and remind you that two reviews and an article are a week overdue and that your agent is baying like a bloodhound for the manuscript you promised him eighteen months previously. Assuring them, quite

Mandrill

untruthfully, that you will be back very shortly, you make your way down to the penguins.

On the way you meet Stefan grinning to himself, and on asking him what the joke is, he tells you that he was in one of the lions' dens, cleaning it out, when, glancing over his shoulder, he was surprised to see a visitor standing there, using the place as a lavatory.

'What are you doing?' inquired Stefan.

'Well, this is the gents', isn't it?' replied the man peevishly.

'No, it isn't. It's the lions' den,' replied Stefan.

Never had an exit been so rapidly performed from a public convenience, he told me.

Having worked out a complicated but very beautiful plan for the penguin pool, you then have to work out an equally complicated and beautiful plan for getting the scheme passed by Catha, the administrative secretary, who holds the Zoo's purse-strings in a grip so firm it requires as much ingenuity to prise money out of it as it would do to extract a coin from a Scotsman's sporran. You march to the office, hoping to find her in a sunny, reckless mood, instead of which she is glowering over an enormous pile of ledgers. Before you can start extolling the virtues of your penguin pond idea, she fixes you with a gimlet green eye and in a voice like a honey-covered razor-blade informs you that your last brilliant idea came to approximately twice what you had estimated. You express bewilderment at this and gaze suspiciously at the ledger, implying, without saying so, that her addition must be wrong. She obligingly does the sum in front of you, so that there will be no argument. Feeling that this precise moment is not perhaps the best one to broach the subject of the penguin pond, you back hastily out of the office and go back into the Zoo.

You are spending a pleasant ten minutes making love to the woolly monkeys through the wire of their cage, when suddenly your secretary materializes at your elbow in the most unnerving fashion, and before you can think up a suitable excuse she has reminded you once more about the reviews, the article and the book, and has dragged you disconsolately back to your office.

As you sit there racking your brains to try to think of something tactful to say about a particularly revolting book that has been sent to you for review, a constant procession of people appears to distract your attention. Catha comes in with the minutes of the last meeting, closely followed by Les who wants to know what mesh of wire to put on the new cage. He is followed by John who wants to know if the mealworms have arrived, as he is running short, and then Jeremy appears to tell you that the dingoes have just had eleven pups. I defy any writer to write a good review when his mind is occupied with the insoluble problem of what to do with eleven dingo pups.

Eventually, you manage to finish the review and slip once more into the Zoo. It is getting towards evening now and the crowds are thinning out, drifting away up the main drive to the car park, to wait for their buses or coaches. The slanting rays of the sun floodlight the cage in which the crowned pigeons live: giant powder-blue birds with scarlet eyes and a quivering crest of feathers as fine as maidenhair fern. In the warmth of the setting sun they are displaying to each other, raising their maroon-coloured wings over their backs, like tombstone angels, bowing and pirouetting to one another and then uttering their strange, booming cries. The chimpanzees are starting to scream peevishly, because it is nearing the time for their evening milk, but they pause in their hysterical duet to utter greeting to you as you pass.

Up in the Small Mammal House the night creatures are

starting to come to life, creatures that all day long have been nothing but gently snoring bundles of fur: bushbabies with their enormous, perpetually horrified eyes, creep out of their straw beds and start to bound about their cages, silently as thistledown, occasionally stopping by a plate to stuff a handful of writhing mealworms into their mouths; pottoes, looking like miniature teddy-bears, prowl about the branches of their cage, wearing guilty, furtive expressions, as though they were a convention of cat burglars; the hairy armadillo, you are relieved to see, has roused itself out of its stupor and is now the right way up, puttering to and fro like a clockwork toy.

Downstairs, with growls of satisfaction, the gorillas are receiving their milk. Nandy likes to drink hers lying on her stomach, sipping it daintily from a stainless steel dish. N'Pongo has no use for this feminine nonsense and takes his straight from the bottle, holding it carefully in his great black hands. He likes to drink his milk sitting up on the perch, staring at the end of the bottle with intense concentration. Jeremy has to stand guard, for when N'Pongo has drained the last dregs he will just simply open his hands and let the bottle drop, to shatter on the cement floor. All round, the monkeys

Dingo puppies

are gloating over their evening ration of bread and milk, uttering muffled cries of delight as they stuff their mouths and the milk runs down their chins.

Walking up towards the main gate you hear the loud ringing cries of the sarus cranes: tall, elegant grey birds with heads and necks the colour of faded red velvet. They are performing their graceful courting dance in the last rays of the sun, against a background of blue and mauve hydrangea. One will pick up a twig or a tuft of grass, and then, with wings held high, will twirl and leap with it, tossing it into the air and prancing on its long slender legs, while the other one watches it and bows as if in approval. The owls are now showing

Bushbaby

Potto

signs of animation. Woody, the Woodford's owl, clicks his beak reprovingly at you, as you peer into his cage, and over his immense eyes he lowers blue lids with sweeping eyelashes that would be the envy of any film star. The white-faced Scops owls that have spent all day pretending to be grey, decaying tree stumps, now open large, golden eyes and peer at you indignantly.

Shadows are creeping over the flower-beds and the rockery. The peacock, as exhausted as an actor at the end of a long run, passes slowly towards the walled garden, dragging his burnished tail behind him and leading his vacant-eyed harem towards their roosting place. Sitting on top of the granite cross that surmounts the great arch leading into the courtyard, is our resident robin. He has a nest in a crevice of the wall, half hidden under a waterfall of blue-flowered rock plants. So, as his wife warms her four eggs, he sits on top of the cross and sings his heart out, gazing raptly at the western sky where the setting sun has woven a sunset of gold and green and blue.

As the light fades, the robin eventually ceases to sing and flies off to roost in the mimosa tree. All the day noises have now ceased and there is a short period of quietness before the night cries take over. It is started inevitably by the owls – beak-clicking and a noise like tearing calico from the white-faced Scops owls, a long, tremulous and surprised hoot from the Woodford's owl, and a harsh, jeering scream from the Canadian horned owls. Once the owls have started, they are generally followed by the Andean fox who sits forlornly in the centre of his cage, throws back his head and yaps shrilly at the stars. This sets off the dingoes in the next cage and they utter a series of gently melodious howls that are so weird and so mournful they make you want to burst into tears. Not to be outdone, the lions take up the song – deep, rasping, full-throated roars that tail off into a satisfied gurgle that sounds unpleasantly as though they have just found a hole in the wire.

In the Reptile House snakes that had been so lethargic all day now slide round their cages, bright-eyed, eager, their tongues flicking as they explore every nook and cranny for food. The geckoes, with enormous, golden eyes, hang upside down on the roof of their cage, or else with infinite caution stalk a dishful of writhing mealworms. The tiny, yellow and black corroboree frogs (striped like bulls' eyes and the size of a cigarette butt) periodically burst into song: thin, reedy piping that has a metallic quality about it, as if someone were tapping a stone with a tiny hammer. Then they relapse into silence and gaze mournfully at the ever-circling crowd of fruit flies that live in their cage and form part of their diet.

Outside, the lions, the dingoes and the fox are quiet; the owls keep up their questioning cries. There is a sudden chorus of hysterical screams from the chimpanzees' bedroom and you

Woodford's owl

know that they are quarrelling about who should have the straw.

In the Mammal House the gorillas are now asleep, lying side by side on their shelf, pillowing their heads on their arms. They screw up their eyes in your torch beam and utter faint indignant growls that you should disturb them. Next door, the orang-utans, locked passionately in each other's arms, snore so loudly that it seems as though the very floor vibrates. In all the cages there is deep relaxed breathing from sleeping monkeys, and the only sound apart from this is the steady patter of claws as the nine-banded armadillo, who always seems to suffer from insomnia, trots about his cage, making and remaking his bed, carefully gathering all the straw into one corner, smoothing it down, lying on it to test its comfort, and then deciding that the corner is not a suitable one for a bedroom, removing all the bedding to the opposite end of the cage and starting all over again.

Upstairs, the flying squirrels gaze at you with enormous, liquid eyes, squatting fatly on their haunches, while stuffing food into their mouths with their delicate little hands. Most of the parrots are asleep, but Suku, the African grey, is incurably inquisitive, and, as you pass, never fails to pull his head out from under his wing and watch to see what you are doing. As you make your departure, he shuffles his feathers – a whispering silken noise – and then in a deep, rather bronchial voice says: 'Good night, Suku,' to himself in tones of great affection.

As you lie in bed, watching through the window the moon disentangling itself from the tree silhouettes, you hear the dingoes starting again their plaintive, flute-like chorus, and then the lions cough into action. Soon it will be dawn and the chorus of bird-song will take over and make the cold air of the morning ring with song.

CHAPTER TWO

A PORCUPINE IN THE PARISH

Dear Mr Durrell,
I would like to join one of your expeditions. Here are my qualifica-
tions and faults:
*   36 years old, single, good health, a sport, understand children*
and animals, except snakes; devoted, reliable, excellent; young in
character. My hobbies are playing the flute, photography and writing
stories. My nerves are not too steady; am disagreeable if anybody
insults my country or my religion (Catholic). In the event of my
accompanying you, it would be everything paid – on the other hand, if
you are a snob and you don't mean what you write, I regret to say I
do not wish to know you. Hoping to hear from you soon . . .

I SOON found, to my relief, that Jersey appeared to have
taken us to its heart. The kindness that has been shown to
us during the five years of our existence is tremendous, both
from officials and from the Islanders themselves. After all,
when living on an island eight miles by twelve you may be
pardoned for having certain qualms when someone wants to
start a zoo and import a lot of apparently dangerous animals.
You have vivid mental pictures of an escaped tiger stalking
your pedigree herd of Jersey cows, of flocks of huge and
savage deer browsing happily through your fields of daffodils,
and gigantic eagles and vultures swooping down on your
defenceless chickens. I have no doubt that a lot of people
thought this, especially our nearest neighbours to the Manor,
but nevertheless they welcomed us without displaying any
symptoms of unease.

In a zoo of five or six hundred animals the variety and
quantity of food they consume is staggering. It is the one
thing that must not be stinted if they are to be kept healthy

and happy; and, above all, the food must not only be plentiful, but good. Cleanliness and good food go a very long way to cutting down disease. A creature that is well fed and kept in clean surroundings has, in my opinion, an eighty per cent better chance of escaping disease, or, if it contracts a disease, of recovering. Unfortunately, a great many people (including, I am afraid, some zoos) still suffer from the extraordinary delusion that anything edible but not fit for human consumption is ideal for animals. When you consider that, in the wild state, most animals – unless they are natural carrion feeders – always eat the freshest of food, such as fresh fruit and freshly killed meat, it is scarcely to be wondered at when they sicken and die if fed on a diet that is 'not fit for human consumption'. Of course, in all zoos a lot of such food *is* fed, but in most cases there is nothing at all wrong with it. For example, a greengrocer opens a crate of bananas and finds that many of the fruit have black specks or blotches on the skin. There is nothing wrong with the fruit, but his customers demand yellow bananas, and will not buy discoloured ones. If a zoo did not buy it the fruit would be wasted. Sometimes the greengrocer has fruit or vegetables which have reached that point of ripeness where another twenty-four hours in the shop and the whole lot will have to be thrown away. In that case they are sold to a zoo that can use them up rapidly.

Some time ago a greengrocer telephoned us, inquiring if we would like some peaches. He explained that his deep-freeze had gone wrong and that it contained some South African peaches which had gone black just round the seed. There was absolutely nothing wrong with them, he reassured us, but they were unsaleable. We said we would be delighted to have them, thinking that a couple of crates of peaches would be a treat for some of the animals. A few hours later, a huge lorry rolled into the grounds, stacked high with boxes. There must have been anything up to thirty or forty, and the financial loss this represented to the greengrocer must have been staggering. They were some of the largest and most succulent peaches I have ever seen; we tipped cratefuls of them into the cages, and the animals had a field day. Within half an hour all the monkeys were dripping peach juice and could hardly move; several members of the staff, too, were surreptitiously wiping peach juice off their chins. There was, as I say, nothing wrong with the peaches: they were just unsaleable. But it might happen that someone else, in the most kindly way, would bring us a whole lorry load of completely rotten and mildewed peaches, and be hurt and puzzled when we refused them on the grounds that they were unfit for animal consumption. One of the biggest killers in a zoo is that rather nebulous thing called enteritis, an infection of the stomach. This, in itself, can cause an animal's death, but even if it is only a mild attack it can weaken the creature, and thus open the door to pneumonia or some other deadly complaint. Bad fruit can cause enteritis quicker than most things; thus care must be taken over the quality of the fruit fed to the animals.

As soon as the people of Jersey knew what our requirements were in the matter of food, they rallied round in the most extraordinarily generous way. Take the question of calves, for instance. In Jersey most of the bull-calves are slaughtered at birth and, until we arrived, they were simply buried, for they were too small to be marketable. We discovered this quite by accident, when a farmer telephoned us and asked, rather doubtfully, if a dead calf was any use to us. We said we would be delighted to have it, and when he brought it round he asked us if we would like any more. It was then that we found out there was this wonderful source of fresh meat: meat which – from the animal point of view – could not have been more natural, for not only was it freshly killed (sometimes still warm), but they could also have the hearts, livers and other internal organs which were so good for them. Gradually, the news spread among the farmers, and before long – at certain times of the year – we were receiving as many as sixteen calves a day, and farmers were travelling from one end of the Island to the other, delivering them to us. Other farmers, not to be outdone, offered us tomatoes and apples, and would bring whole lorry-loads round, or let us go to collect as many as we could take away. One man telephoned to say he had a 'few' sunflowers, the heads of which were now ripe, would we like them? As usual we said yes, and he turned up in a small open truck piled high with gigantic sunflower heads, so that the whole thing looked like a sun chariot. The heads were not fully ripe, which meant that the kernel of each seed was soft and milky; we simply cut up the heads as if they were plum cakes and put big slices in with such creatures as the squirrels, mongooses and birds. They all went crazy about the soft seeds and simply gorged themselves.

But these are all the more normal types of food. In a zoo you can use a great many very unusual items of diet, and in acquiring these we were again helped by the local people. There was one elderly lady who used to cycle up to the Zoo once or twice a week on an antediluvian bicycle and spend the afternoon talking to the animals. Whenever she saw me she would back me into a corner and for half an hour or so tell me what tricks her favourite animals had been up to that day. She was, I discovered, a lavatory attendant in St Helier. One day I happened to meet her when I had been out collecting some acorns for the squirrels. She watched entranced while the squirrels sat up on their hind legs, twirling the acorns round and round in their paws as they chewed them. She then told me that she knew of a great many churchyards in which grew fine oak trees, and vowed that she would, herself, bring some acorns for the squirrels at the end of the week. Sure enough, on the Sunday she appeared, pedalling strenuously up to the

Zoo on her ancient bicycle, the front basket of which was filled to the brim with plump acorns, and there was another large carrier-bagful strapped – somewhat insecurely – to the back of her vehicle. Thereafter, she used to bring us a regular supply of acorns every week, until the squirrels became quite blasé about them and even started to store them in their beds.

Another food item for which we are always grateful is what could be loosely called 'live food', that is to say earwigs, woodlice, grasshoppers, moths and snails. Here a great many people come to our rescue, and they turn up at the Zoo with jam-jars full of woodlice and other creatures, and biscuit-tins full of snails, of which they are, of course, only too glad to see the last. The earwigs, woodlice and so on are fed to the smaller reptiles, the amphibians and some of the birds. The snails we feed to the larger lizards, who scrunch them up with avidity, eating shell and all as a rule.

In order to pad out the animals that I had brought back from West Africa and South America, we had, of course, to acquire from different sources several other creatures. The most amusing of these was, undoubtedly, the bird I mentioned before, Trumpy, the trumpeter. Not only had he appointed himself the Zoo's clown but also the Zoo's settler-in. As soon as we got a new creature, Trumpy managed to hear of it, and would come bouncing along, cackling to himself, to settle it in. He would then spend twenty-four hours standing by the cage (or preferably in it, if he could) until he thought that the new arrival was firmly established, whereupon he would bounce back to his special beat in the Mammal House. Sometimes Trumpy's settling efforts were on the risky side, but he seemed to be too dim-witted to realize the danger. When Juan and Juanita, the white-collared peccaries, were first released into their paddock there was Trumpy to settle them in. The pigs did not seem to mind in the slightest, and so Trumpy did his twenty-four-hour stint and departed. But later on, when Juan and Juanita had just had their first litter, and had brought them out into the paddock for the first time, Trumpy flew gaily over the fence to settle in the babies. Now, Juan and Juanita had not minded this for themselves, but they thought that Trumpy's efforts on behalf of their piglets held some hidden menace. They converged on Trumpy (who was standing on one leg and eyeing the piglets benignly), their fur bristling, their tusks clattering like castanets. Trumpy woke out of his trance with a start, and only a skilful bit of dodging and a wild leap saved him. It was the last time he attempted to go into the peccary paddock. When we dammed up the little stream in the sunken water-meadow and constructed a small lake for the black-necked and coscoroba swans I had brought

back from South America, Trumpy was there to supervise the work, and when the swans were eventually released he insisted, in spite of all our entreaties, on standing up to his ankles in water for twenty-four hours to settle them in. It did not appear to have any effect on the swans, but Trumpy enjoyed it.

Another new acquisition was the fine young male mandrill, Frisky. With his blue and red behind, and his blue and red nose, Frisky was a fine sight. If you went near his cage he would peer at you with his bright, amber-coloured eyes, lift his eyebrows up and down, as if in astonishment, and then, uttering throaty little grunts, he would turn round and present

his backside to you, peering over his shoulder to see what effect his sunset-rear was achieving. Frisky was, of course, exceedingly inquisitive, like all members of his family, and one bright spring day this was his undoing. We were having the tops of the monkey cages repainted in a pleasant shade of mushroom, and Frisky had been watching this operation with the keenest interest. He was obviously under the impression that the paint pot contained some delicious substance, probably like milk, which would repay investigation. He had not had a chance to find out, however, for the painter, in the most selfish and boorish manner, had kept the paint pot close beside him. But patience is always rewarded, and after a few hours Frisky had his chance. The painter left the pot unguarded while he went to fetch something, and Frisky seized the opportunity. He pushed his arm through the wire, grabbed the edge of the paint pot and pulled. The next moment he was spluttering and choking under a waterfall of mushroom-coloured paint, and almost instantly, he discovered, he had turned into a mushroom-coloured mandrill. There was really not much that we could do, for you cannot take a half-grown mandrill out of its cage and wash it as though it were a poodle. However, when the paint had dried as hard as armour on his fur, he looked so miserable that we decided to put him into the next door cage, which contained a female baboon and two female drills, in the hope that they would clean him. When Frisky was let in with them, they viewed him with alarm, and it was some time before they plucked up enough courage to approach him. When they did, however, and found out what was the matter with him, they gathered round enthusiastically and set about the task of giving Frisky a wash and brush up. The trouble was that the paint had dried so hard on to the fur that the three females had to use a great deal of force, and so, although at the end of two days they had removed all the paint, they had also removed a vast amount of Frisky's fur with it. Now, instead of a mushroom-coloured mandrill, we had a partially bald and slightly shame-faced looking mandrill.

Another newcomer was our lion, who went under the time-honoured name of Leo. He was one of the famous Dublin Zoo lions, and was probably about the fiftieth generation born in captivity. On his arrival he was only about the size of a small dog, and so he was housed in a cage in the Mammal House, but he grew at such a pace that it was soon imperative that we find him more spacious quarters. We had just finished construction on a large cage for the chimpanzees, and decided we would put Leo in that until we could get around to building him a cage of his own. So Leo was transferred, and settled down very happily. I was glad to see, when his mane started to develop, that he was going to be a blond lion, for in my experience the lions with blond manes, as opposed to dark manes, have always nice, if slightly imbecile, characters. This theory has been amply born out by Leo's behaviour. He had in his cage a large log as a plaything, and a big, black rubber bucket in which he received his water ration. This bucket fascinated him, and after he had drunk his fill he would upset the remains of the water and then pat the bucket with his great paws, making it roll round the cage so that he could stalk it and pounce on it. One day I was in the grounds when a lady stopped me to inquire whether we had acquired Leo from a circus. Slightly puzzled, I said, 'No,' and asked her why she should think so. 'Because,' she replied, 'he was doing such clever tricks.' I discovered that he had, by some extraordinary means, managed to wedge the rubber bucket on his head, and was walking round and round the cage proudly, wearing it like a hat.

In his second year Leo decided, after mature reflection, that it was a lion's duty to roar. He was not awfully sure how to go about it, so he would retire to quiet corners of his cage and

practise very softly to himself, for he was rather shy of this new accomplishment and would stop immediately and pretend it was nothing to do with him if you came in view. When he was satisfied that the timbre was right and his breath control perfect, he treated us to his first concert. It was a wonderful moonlight night when he started, and we were all delighted that Leo was, at last, a proper lion. A lion roaring sounds just like someone sawing wood on a gigantic, echoing barrel. The first coughs or rasps are quick and fairly close together, and you can imagine the saw biting into the wood; then the coughs slow down and become more drawn out, and suddenly stop, and you instinctively wait to hear the thud of the sawn-off piece hitting the ground. The trouble was that Leo was so proud of his accomplishment that he could not wait until nightfall to give us the benefit of his vocal cords. He started roaring earlier and earlier each evening, and would keep it up solidly all night, with five-minute intervals for meditation in between each roar. Sometimes, when he was in particularly good voice, you could imagine that he was sitting on the end of your bed, serenading you. We all began to be somewhat jaded. We found that if we opened the bedroom window and shouted, 'Leo, *shut up*,' this had the effect of silencing him for half an hour; but at the end of that time he would decide that you had not really meant it and would start all over again. It was a very trying time for all concerned. Now, however, Leo has learnt to roar with a certain amount of discretion, but even so there are nights – especially at full moon – when the only thing to do is to put the pillow over your head and curse the day you ever decided you wanted a zoo.

We also obtained in our first year two South African penguins, called Dilly and Dally. I hasten to add that they were not christened by us, but arrived with these revolting names stencilled on their crate. We had prepared a pool for them in the shade of some trees bordering the main drive, and here they seemed quite content. Trumpy, of course, spent twenty-four hours in their pen with them, and seemed faintly disgruntled that the pool was too deep for him to join Dilly and Dally in it. After settling them in, he took a great fancy to the penguins and paid them a visit every morning, when he would stand outside the wire, making his curious booming cry, while Dilly and Dally would point their beaks skywards and bray to the heavens, like a couple of demented donkeys.

I am not quite sure when the rift in this happy friendship appeared, or for what reason, but one morning we saw Trumpy fly over into the penguin enclosure and proceed to beat up Dilly and Dally in the most ferocious manner. He flew at them, wings out, feathers bristling, pecking and scratching, until the two penguins (who were twice his size) were forced to take refuge in the pool. Trumpy stood on the edge of the pond and cackled triumphantly at them. We chased Trumpy out of the enclosure and scolded him, where-upon he shuffled his feathers carelessly and stalked off non-chalantly. After that we had to watch him, for he took advantage of every opportunity to fly over the wire and attack poor Dilly and Dally who, at the sight of him, would flop hysterically into the water. One morning he did this once too often. He must have flown over very early, before anyone was about, intent on giving Dilly and Dally a bashing, but the penguins had grown tired of these constant assaults, and rounded on him. One of them, with a lucky peck, must have caught him off balance and knocked him into the pool, from which – with his waterlogged feathers – he could not climb out. This was the penguins' triumph, and as Trumpy floun-dered helplessly, they circled round, pecking at him viciously with their razor-sharp beaks. When he was found, he was still floating in the pond, bleeding profusely from a number of pecks, and with just enough strength to keep his head above water. We rushed him into the house, dried him and anointed his wounds, but he was a very sick and exhausted bird, and black depression settled on the Zoo, for we all thought he would die. The next day there was no change, and I felt it was touch and go. As I was sipping my early morning tea on the third day, I suddenly heard, to my amazement, a familiar thrumming cry. I slipped out of bed and looked out of the window. There, by the lavender hedge in the courtyard, was Trumpy, a slightly battered and tattered trumpeter who limped a little, but still with the same regal air of being the owner of the property. I saluted him out of the window, and he cocked a bright eye at me. Then he shuffled his torn feathering to adjust it to his liking, gave his loud, cackling laugh and stalked off towards his beat in the Mammal House.

Another new arrival that caused us a certain amount of trouble, one way and another, was Delilah. She was a large female African crested porcupine, and she arrived up at the airport in a crate that looked suitable for a couple of rhino-ceros. Why she had been crated like this became obvious when we peered into the crate, for even in that short air journey she had succeeded in nearly demolishing one side with her great yellow teeth. When she saw us looking into the crate, she uttered a series of such fearsome roars and gurks that one would have been pardoned for thinking it contained a pride of starving lions. She stamped her feet petulantly on the floor of the crate, and rattled and clattered her long black and white quills like a crackle of musketry. It was quite obvious that Delilah was going to be a personality to be reckoned with.

On our return to the Zoo we had to chivy her out of her rapidly disintegrating crate and into a temporary cage, while her permanent home was under construction. During this process she endeared herself to at least one member of the staff by backing sharply into his legs. The experience of having several hundred extremely sharp porcupine quills stabbed into your shins is not exactly an exhilarating one. By the time Delilah was installed in her temporary home there were several more casualties, and the ground was littered with quills, for Delilah, like all porcupines, shed her quills with gay abandon at the slightest provocation.

The old fable of a porcupine being able to shoot its quills out like arrows is quite untrue. What actually happens is this. The quills, some of them fourteen inches long, are planted very loosely in the skin of the back. When the animal is harried by an enemy, what it does it to back rapidly into the adversary (for all the quills point backwards), jab the quills into him as deeply as possible, and then rush forward again. This action not only drives the quills into the enemy, but pulls them loose from the porcupine's skin, so the enemy is left looking like a weird sort of pincushion. This action is performed so rapidly that, in the heat of battle, as it were, you are quite apt to get the impression that the porcupine *has* shot its adversary full of quills. This delightful action Delilah used to indulge in with great frequency, and, therefore, at feeding and cleaning times you had to be prepared to drop everything and leap high and wide at a moment's notice.

Porcupines are, of course, rodents, and the giant crested species – since it spreads from Africa into parts of Europe – has the distinction of being the largest· European rodent, bigger even than the beaver. It is also the largest of the porcupines, for, although there are many different species scattered about the world, none of them comes anywhere near the size of the crested one. In North and South America the porcupines are, to a large extent, arboreal, and the South American kind even have prehensile tails to assist them in climbing. The other porcupines found in Africa and Asia are rather small, terrestrial species, that generally have fairly long tails ending in a bunch of soft spines like the head of a brush, and this they rattle vigorously in moments of stress. Without doubt, as well as being the biggest, the great crested porcupine is the most impressive and handsome member of the family.

It was not long before we had Delilah's new home ready, and then came the great day on which we had to transport her to it from one end of the Zoo to the other. We had learnt from bitter experience that trying to chivy Delilah into a crate was worse than useless. She simply put up all her spines, gurked at us fiercely and backed into everything in sight, parting with great handfuls of quills with a generosity I have rarely seen equalled. The mere sight of a crate would send her off into an orgy of foot-stamping and quill-rattling. We had learnt that there was only one way to cope with her: to let her out of the cage and then two people, armed with brooms, to chivy her along gently. Delilah would stride out like one of the more muscular and prickly female Soviet athletes, and as long as you kept her on a fairly even course by light taps from the brushes you could keep her going for any distance.

This was the method we decided to employ to transfer her to her new quarters, and to begin with all went well. She started off at a great lick down the main drive, Jeremy and I panting behind with our brushes. We successfully made her round the corner into the courtyard, but once there a suspicion entered her head that she might be doing exactly what we wanted her to do. Feeling that the honour of the rodents was at stake, Delilah proceeded to run round and round the courtyard as though it was a circus ring, with Jeremy and me in hot pursuit. Then, whenever she had got us going at a good pace, she would suddenly stop and go into reverse, so that we would have to leap out of the way and use our brushes as protection. After a few minutes of this, there appeared to be more quills sticking in the woodwork of the brushes than there were in Delilah. Eventually, however, she tired of this game, and allowed us to guide her down to her new cage without any further ado.

She lived very happily in her new quarters for about three months before the *wanderlust* seized her. It was a crisp winter's evening when Delilah decided there might be something in the outside world that her cage lacked, and so setting to work with her great curved yellow teeth she ripped a large hole in the thick interlink wire, squeezed her portly form through it and trotted off into the night. It so happened that on that particular evening I had gone out to dinner, so the full honours of the Battle of the Porcupine go to John.

At about midnight my mother was awakened by a car which had driven into the courtyard beneath her bedroom window and was tooting its horn vigorously. Mother, leaning out of the window, saw that it was one of our nearest neighbours from the farm over the hill. He informed Mother that there was a large and, to judge by the noises it was making, ferocious creature stamping about in his yard, and would we like to do something about it. Mother, who always has a tendency to fear the worst, was convinced that it was Leo who had escaped, and she fled to the cottage to wake John. He decided from the description that it must be Delilah, and pausing only for a broom, he leapt into the Zoo van and drove up to the farm. There, sure enough, was Delilah, stamping about in the moonlight, gurking to herself and rattling her quills. John explained to the farmer that the only way to get Delilah back to the Zoo was to brush her, as it were, along the half mile or so of road. The farmer, though obviously thinking the whole procedure rather eccentric, said that if John would undertake that part of it, he would undertake to drive the Zoo van back again.

So John set off, clad in his pyjamas, brushing a snorting, rattling Delilah down the narrow moonlit road. John said he had never felt such a fool in his life, for they met several cars full of late-night revellers, and all these screeched to a halt and watched in open-mouthed astonishment the sight of a man in pyjamas brushing along a plainly reluctant porcupine. Several of them, I am quite sure, must have hurried home to sign the pledge, for after all, the last thing you expect to find wandering about a respectable parish is an infuriated porcupine pursued by a highly embarrassed man in night attire. But at last John brought her safely back to the Zoo and then, to her great indignation, locked her up in the coal cellar. For, as he explained, it had a cement floor and two-foot thick granite walls, and if she could break out of that she deserved her freedom and, as far as he was concerned, she could have it.

Not long afterwards, Delilah caused trouble in quite another context. As the Zoo needs every form of publicity it can obtain, television was clearly one of the best mediums, and so I tried to popularize it by this means whenever possible. A television producer once said to me that if he could produce a programme without a television personality or professional

actor he would be a happy man. I could see his point, but he did not know that there could be something infinitely more harrowing than putting on a programme with a television personality or a professional actor. He had never undertaken one with live wild animals, the difficulties of this making the strutting and fretting of television personalities and actors fade into insignificance. When making a programme with animals, they either behave so badly that you are left a jittering mass of nerves in the end, or else they behave so well that they steal the show. Whichever way it is, you cannot win, and anyone (in my considered opinion) who undertakes to do such a job, should be kindly and firmly conducted by his friends to the nearest mental home. If you let him do the programme, he will end there anyway, so you are merely anticipating.

One of the first programmes I did was devoted to the primates, or monkey family, of which the Zoo boasted a rather fine collection. For the first time, live, on television, I could show the great British public a splendid array of creatures ranging from the tiny, large-eyed bushbabies, through the lorises, the Old and New World monkeys, to the gorilla and chimpanzee, with myself thrown in as an example of *Homo sapiens*. I had no qualms about this: the monkeys and apes were all extremely tame, the bushbabies would be confined in glass-fronted cases, and the lorises would be on upright branches where, I knew, they would simply curl up and sleep until awakened by me during the programme. At least, that is how it should have worked, but unfortunately I had not taken into consideration the effects of the journey, for the Island of Jersey is an hour's flying time from the City of Bristol where the programme was to be recorded. By the time the animals had been crated, flown to Bristol and unloaded in the dressing-room which had been put at their disposal, they were all in a highly neurotic state of mind. So was I.

When the time for the first rehearsal approached, all the

monkeys had to be removed from their travelling crates, have belts and leashes attached to them, and be tethered (one to each compartment) on a construction that resembled a miniature cow stall. The monkeys, hitherto always tame, placid and well-behaved, took one look at the cow stall and had what appeared to be a collective nervous breakdown. They screamed, they bit, they struggled; one broke his leash and disappeared behind some piled scenery, from which he was extracted – yelling loudly and covered with cobwebs – after about half an hour's concentrated effort. Already rehearsal was fifteen minutes overdue. At last we had them all in position and more or less quiet. I apologized to the producer and said that we would be ready in next to no time, for all we had to do was to put the lorises on their respective tree trunks, and this – with such lethargic animals – would be the work of a moment. We opened the cage doors, expecting to have to chivy the sleepy lorises out on to their trees, but instead they stalked out like a couple of racehorses, their eyes blazing with indignation, uttering loud cat-like cries of disgust and warning. Before anyone could do anything sensible, they had rushed down their tree trunks and were roaring across the studio floor, their mouths open, their eyes wide. Technicians departed hurriedly in all directions, except a few of the bolder ones who, with rolled-up newspapers as weapons, endeavoured to prevent the determined lorises from getting among the scenery, as the monkey had done. After further considerable delay we managed to return the lorises to their travelling crates, and the Props Department was hurriedly summoned to attach to the bottom of each tree a cardboard cone that would prevent the creatures from getting a grip and so climbing down on to the floor. Rehearsals were now an hour overdue. At last we were under way, and by this time I was in such a state of nerves that the rehearsal was a shambles: I forgot my lines; I called most of the animals by

the wrong names; the slightest sound made me jump out of my skin, for fear something had escaped, and to cap it all Lulu, the chimp, urinated copiously, loudly and with considerable interest in her own achievement, all over my lap. We all retired to lunch with black circles under our eyes, raging headaches and a grim sense of foreboding. The Producer, with a ghastly smile, said she was sure it would be all right, and I, trying to eat what appeared to be fried sawdust, agreed. We went back to the studio to do the recording.

For some technical reason that defeats me, it is too expensive or too complicated to cut a tele-recording. So it is exactly like doing a live programme: if you make a mistake, it is permanent. This, of course, does not help to bolster your confidence in yourself; when you are co-starring with a number of irritated and uninhibited creatures like monkeys you start going grey round the temples before you even begin. The red light went on, and with shaking hands I took a deep breath, smiled a tremulous smile at the camera, as if I loved it like a brother, and commenced. To my surprise, the monkeys behaved perfectly. My confidence started to return. The bushbabies were wonderful, and I felt a faint ray of hope. We reached the lorises and they were magnificent. My voice lost its tremolo and, I hoped, took on a firm, manly, authoritative note. I was getting into my stride. Just as I was launching myself with enthusiasm into the protective postures of a potto – believe it or not – the Studio Manager came over and told me that there had been a breakdown in the tele-recording and we should have to start all over again.

Of course, after an experience like this, one is mental to even try to do any more television. But I had agreed to do five more. The five I did, I must admit, were not quite as trying as the monkey programme, but some of the highlights still live vividly in my memory, and occasionally I awake screaming in the night and have to be comforted by Jacquie. There was, for example, the programme I did on birds. The idea was to assemble as many different species as possible, and show how their beaks were adapted for their varying ways of life. Two of the birds were to be 'star' turns, because they did things to order. There was, for instance, Dingle the chough. This member of the crow family is extremely rare in Great Britain now, and we are very lucky to have him. They are clad in funereal black feathering, but with scarlet feet and a long, curved scarlet beak. Dingle, who had been hand-reared, was absurdly tame. The second 'star' was a cockatoo named – with incredible originality by its previous owner – 'Cocky'. Now, this creature would, when requested, put up its amazing crest and shout loudly, a most impressive sight. All the other birds taking part in the programme did nothing at all: they were all, very sensibly, content to just sit there and be themselves. So my only problems were Dingle and Cocky, and I had great faith in both of them.

The programme was to open with me standing there, Dingle perched on my wrist, while I talked about him. During rehearsals this worked perfectly, for if you scratch Dingle's head he goes off into a trance-like state, and remains quite still. However, when it came to the actual recording, Dingle decided that he had been scratched enough, and just as the red light went on he launched himself off my wrist and flew up into the rafters of the studio. It took us half an hour with the aid of ladders and bribes in the shape of mealworms, meat and cheese (of which he is inordinately fond) to retrieve him, whereupon he behaved perfectly and sat so still on my wrist that he appeared to be stuffed. All went smoothly until we came to Cocky. Here I made the mistake of telling my audience what to expect, which is the one thing not to do with animals. So, while five million viewers gaped expectantly waiting to see Cocky put up his crest and scream, I made desperate attempts to persuade him to do it. This went on for five soul-searing minutes, while Cocky just sat on his perch as immobile as a museum specimen. In despair I eventually moved on to the next bird, and as I did so Cocky erected his crest and screamed mockingly.

There was the occasion, also, of the programme devoted to reptiles. Here I felt I was on safer ground, for, on the whole, they are fairly lethargic creatures and easily handleable. The programme, however, was a chore for me, as I was just in the middle of a bout of influenza, and my presence in the studio was entirely due to the efforts of my doctor who had pumped me full of the most revolting substances to keep me on my feet for the required time. If you are nervous anyway – which I always am – and your head is buzzing under the influence of various antibiotics, you tend to give a performance closely resembling an early silent film. During the first rehearsals all the technicians realized that I was feeling both lousy and strung up, and so when it came to a break they each took it in turn to back me into a corner and try to restore my morale, with little or no effect. We came to the second rehearsal and I was worse than before. Obviously something had to be done, and somebody was inspired enough to think of the answer. During my discourse on members of the tortoise family I mentioned how the skeleton of the beast was, as it were, welded into the shell. In order to show this more clearly I had a very fine tortoise shell and skeleton to demonstrate. The bottom half of the shell was hinged, like a door, and upon opening it all the mysteries and secrets of the tortoise's anatomy were revealed. Having done my little introduction on the tortoise family, I then opened the underside of the shell and, to my susprise, instead of just finding the skeleton therein, I found a piece of cardboard on which the words 'NO VACANCIES' had been carefully printed. It was a few minutes before order was restored in the studio, but I felt much better, and the rest of the rehearsal went off without a hitch.

Delilah cropped up in a programme which I did on adaptation. I thought she would be a very good example of the way an animal protects itself, and certainly she showed this off to advantage. When we came to put her into the crate, she charged wildly in all directions, backing into us and the woodwork, and leaving spines embedded in the sides of the crate and in the ends of the brushes. She gurked and roared and rattled her quills throughout the trip to Bristol, and the studio hands, who unloaded her on arrival there, were for some considerable time under the impression that I had brought a fully grown leopard with me. Then we had to transfer Delilah from her travelling box and into the special studio cage that had been built for her. This took half an hour and by the time we had achieved it Delilah had stuck so many quills into so much of the studio scenery that I began to wonder whether she would be completely bald for her debut on television. During the actual transmission she behaved, to my amazement, perfectly, doing all the things that I wanted: she gurked fearsomely, she stamped her feet and rattled her quills like castanets, as though she were a born television star. By the end of the show I was feeling quite friendly towards her and beginning to think that I might well have misjudged her. Then came the moment of inducing her out of the studio cage back into her travelling crate. It took eight of us three-quarters of an hour. One stagehand received a sharp stab in the calf of his leg; two pieces of scenery were irretrievably damaged, and the entire set was pierced so full of porcupine quills it looked as though we had been fighting off a Red Indian attack. I was thankful to get a by then quill-less Delilah back to the Zoo and into her own cage again

I suppose that the terrible things that occur tend to live in one's memory more vividly than the pleasant happenings, and so I look back on the television shows I have done with animals rather in the way that one remembers a nasty series of accidents. There is, however, one incident on which I look back with extreme pleasure, and that was the occasion when the B.B.C. wanted our young gorilla N'Pongo to take part in a programme. They even went to the unprecedented lengths of chartering a small plane to fly us over to Bristol. They also sent a camera man to cover the trip with his camera – a timid individual who confessed to me that he did not like flying, as it made him sick. We took off in brilliant sunshine, and almost immediately dived into black cloud filled to capacity with air pockets. N'Pongo, sitting back in his seat, like a seasoned traveller, thoroughly enjoyed everything. He accepted six lumps of barley sugar to counteract the popping in his ears, peered with interest and excitement out of the window, and when the air pockets began, he fetched out the sick-bag and put it on his head. The poor photographer had become progressively greener, while attempting to film N'Pongo's antics, but when he put the bag on his head this reminder acted in a devastating way, and he dived for his own receptacle and treated it in the way for which it was designed.

CHAPTER THREE

THE COLD-BLOODED COHORT

Dear Mr Durrell,
At a garden fête the other day a lizard was found in the ice-cream
container . . .

I KNOW that it is a confession of acute and depraved eccentricity, but nevertheless I must admit that I am very fond of reptiles. They are not, I grant you, over-burdened with intelligence. You do not get the same reaction from them that you would from a mammal, or even a bird, but still I like them. They are bizarre, colourful, and in many cases graceful, so what more could you want?

Now, the majority of people will confess to you (as though it is something quite unique) that they have an 'instinctive' loathing for snakes, and with much eye-rolling and grimacing they will give you many different reasons for their fear, ranging from the sublime ('it's instinctive') to the ridiculous ('they're all sort of slimy'). I have been, at one time or another, bored by so many snake complex admissions that as soon as the subject of reptiles crops up in conversation with anyone, I want to run away and hide. Ask the average person their views on snakes and they will, within the space of ten minutes, talk more nonsense than a brace of politicians.

To begin with, it is not 'natural' for human beings to fear snakes. You might just as well say that they were naturally afraid of being run over by a bus. Most people, however, are convinced they are born with a built-in anti-snake feeling. This can be quite simply disproved by handing a harmless snake to a child who is too young to have had its head filled with a lot of nonsense about these creatures; the child will hold the reptile and play with it quite happily and without a trace of fear. I remember once putting this point to a woman who had been gurgling on about her snake phobia for what seemed like years.

She was most indignant.

'I've never been taught to fear snakes, I've always been like it,' she said, haughtily, and then added in triumph, 'and my mother was like it, too.'

Faced with such logic, what could one reply?

People's fears of snakes seem to be based on a series of misconceptions. The most common one is the conviction that all these creatures are poisonous. In actual fact, the non-poisonous ones outnumber the poisonous ones by about ten to one. Another very popular idea is that these reptiles are slimy to touch, whereas snakes are dry and cold, and feel no different from a pair of snakeskin shoes or a crocodile-skin handbag. Yet people will insist that they cannot touch a snake because of it sliminess, and think nothing of handling a wet cake of soap.

Our Reptile House is fairly small, but we have a pretty good cross-section of reptiles and amphibia on show. I derive a lot of innocent amusement out of going in there when it is crowded and listening to the general public airing its ignorance with an assurance that is breathtaking. For instance, the snake's tongue: this is purely a scent organ with which the creature smells – hence the way it is flicked rapidly in and out of the mouth; it is also used as a feeler, in the same way that a cat uses its whiskers. The snake experts, however, who visit the Reptile House, know better.

'Cor, Em,' they'll shout excitedly, 'come and look at this snake's sting . . . coo, wouldn't like to be stung by *that*.'

And Em will hurry over and peer horrified at the innocent grass snake, and then give a delicious shudder. All reptiles can, of course, spend long periods completely immobile, when even their breathing is difficult to detect, unless you look closely. The classic remark was delivered by a man who, having peered into several cages in which the reptiles lay unmoving, turned to his wife with an air of one who has been swindled, and hissed:

'They're *stuffed*, Milly.'

A snake moving along the ground or through the branches of a tree is one of the most graceful sights in the world, and when you consider that the creature is walking with its ribs the whole thing becomes even more remarkable. If you watch a moving snake carefully, you can sometimes see the ribs moving beneath the skin as the snake draws itself along. The creature's unblinking stare (another thing to which people object) is not due to the fact that the snake is trying to hypnotize you, but simply that it has no eyelids. The eye is covered with a fine, transparent scale, like a watch-glass. This is very clearly noticed when a snake sheds its skin, which they all do periodically. The skin comes loose around the lips, and then, by rubbing itself against rocks or branches, the snake gradually peels off its old skin. If you examine this shed skin, you can see that the eye scales have been shed as well.

All snakes are adapted for feeding in the same way, but their methods of obtaining their food vary. The non-poisonous ones and the constrictors (like the pythons) grab their prey with their mouths, and then try to throw two or three coils of their body round the victims as rapidly as possible, thus holding and crushing at the same time. The poisonous one, on the other hand, bites its victim and then waits for the poison to take effect, which is generally very soon. Once the prey has undergone its last convulsions, then it can be eaten. The poison fangs, of course, are in the upper jaw, and usually near the front of the mouth. When not in use, they fold back against the gum, like the blade of a pen-knife; as the snake opens its mouth to strike, they drop down into position. The fangs are hollow, similar to a hypodermic needle, or else they have a deep groove running down the back. The poison sac, to which they are connected, lies above the gum. As the snake bites, the poison is forced out and trickles down the groove or hollow in the fang and so into the wound. However, whatever the method of attack, once the prey is dead, the swallowing process is the same in all snakes. The lower jaw is jointed to the upper one in such a way that it can be dislocated at will, and, of course, the skin of the mouth, throat and body is extremely elastic, and so this allows the snake to swallow a creature considerably larger than its own head. Once the food is in the stomach, the slow process of digestion starts. Any portions of the animal that are impossible to assimilate, such as hair, are regurgitated in the form of pellets at a later date. On one occasion a large python was killed, and in its stomach were found four round balls of hair, the size of tennis balls and very hard. On being cut open, each one was found to contain the hoof of a wild pig. These sharp hooves could have damaged the lining of the python's stomach, and so each one had been carefully covered with a thick smooth layer of hair.

In the majority of zoos nowadays they feed dead creatures to the snakes. This is not because it is better for the snakes, or that they prefer it, but simply due to misplaced kindness on the part of the general public, who imagine that a white rat or a rabbit suffers terribly when put into a cage with a snake. That this is nonsense I have proved to my complete satisfaction, for I have seen, in a Continental zoo, a rabbit perched on the back of a python (obviously not hungry), cleaning its whiskers with tremendous sang-froid. The Director of the zoo told me that if white rats were fed to the snakes, it was imperative they should be removed if they were not eaten straight away, otherwise they proceeded to gnaw holes in the snake's body.

While snakes are passive and rather expressionless beasts, lizards can display considerable intelligence and character. One such reptile we had was a mastigure, which I christened Dandy, owing to his great partiality for dandelion flowers. One must, I think, face the fact that mastigures are not the most attractive of lizards, and Dandy was a particularly unattractive member of his species. Nevertheless, his eager personality made him a likeable creature. He had a blunt, rounded head; a fat, flattened body; and a heavy tail covered with short, sharp spikes. His neck was rather long and thin, and this made him look as though he had been put together out of bits of two totally unrelated species. His colour could only be described as a rich, dirty brown. Dandy, as I say, had a liking for dandelion flowers, which amounted to an obsession. He had only to see you approaching the Reptile House with something yellow in your hands, and he would immediately rush to the front of his cage and scrabble wildly against the glass. If it was a dandelion flower you were carrying, you had only to slide back the glass front of his cage and he would gallop out on to your arm, panting with emotion; and then, closing his eyes, he would stretch out his long neck and, like a child waiting to have a sweet popped into its mouth, would open his jaws. If you pushed the flower into his mouth, he would munch away in ecstasy, the petals dangling outside his mouth and making him look as though he had a bright yellow military moustache. He was the only lizard I have known that would genuinely play with you. If he was lying on the sand, and you let your hand creep slowly towards him, as though you were stalking him, he would watch you, his eyes bright, his head on one side. As soon as you were close enough, he would suddenly whip his tail round, give you a gentle bang on the hand with it, and then scuttle away to a new position, and you were then expected to repeat the whole performance. That this was real play I have no doubt, for the blows he

66

dealt you with his tail were very gentle, whereas I have seen him bash other lizards with it and not only send them flying, but draw blood.

Not long after we had Dandy, we had trouble with tegu-exins. These are large, handsome and very intelligent lizards from South America. They can grow to about three and a half feet in length, and their skin is very beautifully patterned in yellow and black. They are very quick-witted, belligerent creatures, and the female we had was quite the most vicious in the Reptile House. Tegus, as they are called for short, have three methods of attack, all of which they employ – together or separately – very cheerfully and without any provocation at all. They will either bite, scratch with their well developed claws, or lash you with their tails. Our female preferred to start hostilities with her tail. As you opened her cage she would regard you with obvious dislike and mistrust, inflate her throat and start to hiss, and at the same time curve her body into a half-moon shape like a bow. Once your hand came near enough, she would suddenly straighten out, and

her tail would lash round like a stockwhip. If she found that this method of defence did not deter you, she would run forward and try to grab you with her mouth. If she succeeded, she would hang on with the tenacity of a bulldog, at the same time bringing up reinforcements in the shape of her sharp, curved hind claws, which could tear chunks off you. I did not think this tegu's character was an exception. After a fair amount of experience with tegus in their natural state, I had come to the conclusion that they were by far the most evilly disposed of the lizards, and were, moreover, so fast and intelligent that they were a force to be reckoned with when in captivity. We were always suffering at the hands, or rather the tail of our female tegu, and so it was with somewhat mixed feelings that we discovered her lying dead in her cage one morning. I was puzzled by her sudden death, for she had appeared to be in the very pink of fighting condition, having bitten me vigorously only a couple of days previously. So I decided to do a rough post-mortem and try to find a clue as to the cause. To my astonishment, on opening the stomach, I found a huge mass of whitish substance, not unlike soft fish roe, which I took to be a gigantic growth of some sort. Wanting to find out more about this mysterious growth, I shipped the body off for a more detailed and expert post-mortem, and awaited the results with interest. When they came through, they were terse and to the point: the mass of white substance had not been a growth, but a large quantity of pure fat. The lizard had died of heart trouble brought on by this fatty condition, and it was suggested that we fed less abundantly in future. On reflection, it was plain, for in the wild state tegus are very active creatures. Therefore, if you confine them in a limited area and give them a rich and continuous food supply, they are bound to become over-fat. I vowed that the next tegus we obtained would be treated very differently.

Our chance came not long afterwards, when a dealer offered us a pair. On arrival they turned out to be wonderful specimens, well marked and with glossy skins: the male with a great, heavy head and fleshy jowls; the female with a longer, more slender head. Contrary to our expectations, they did not prove to be typical tegus at all. Instead of being fierce and unhandleable, they were as tame as kittens, and liked nothing so much as to lie in your arms, being gently rocked, and drowse off to sleep. If you went and stood by their cage, they would make the most frantic and flattering efforts to climb through the glass and into your arms. Apart from these bursts of social activity, they showed little desire to do anything very much, except to lie around in abandoned attitudes, gazing benignly at any human beings who happened to be around in the Reptile House. As a result of all this feverish activity, of course, they grew fatter and fatter, and, viewing their increasing girth with alarm, we decided that something would have to be done, or we would have another couple of heart failures on our hands. The answer was exercise; so, every morning, John would let them both out to wander round the Reptile House, while he did his work. To begin with – for the first two or three days – this worked like a charm, and the tegus, breathing heavily, pottered about the Reptile House floor for a couple of hours each morning. Then, however, they discovered that by climbing over a low barrier they could get into the tortoise pen, over which hung an infra-red light. So, each morning when they were let out, they would rush short-windedly over to the tortoise pen, climb in and settle themselves under the infra-red light and go to sleep. The only answer to this was to cut down on their food, and consequently they were dieted as rigorously as a couple of dowager duchesses at a health resort. Needless to say, they took a very poor view of this, and would gaze plaintively through the glass as they watched the other inmates of the Reptile

House enjoying such delicacies as raw egg, mincemeat, dead rats and chopped fruit. We hardened our hearts, though, and continued with the diet, and within a very short time they had regained their sylph-like figures, and were much more active as a result. Now we let them eat what they like, but at the least sign of corpulence back they will go to the diet until their size is respectable again.

The one Reptile House inhabitant that never seemed to become overweight, no matter how much he ate, was our dragon, known as George. He was a Guiana dragon, a rather rare and interesting kind of lizard from the northern parts of South America. They measure about two feet six inches in length, and have large, heavy heads with big, dark, intelligent eyes. The body and tail are very crocodile-like in appearance, the tail being heavily armoured and flattened on top, whereas the back is covered with heavy scales which are bean-shaped and protrude above the surface of the skin. The colouring is a warm rusty brown, fading to yellowish on the face. They are slow, thoughtful and attractive lizards, and George had a very mild and likeable character.

Probably one of the most remarkable things about Guiana dragons is their feeding habits. Before George arrived we had read up all we could on the species, but none of the textbooks was very helpful. However, they seemed to be perfectly normal lizard-type creatures, so we thought that their diet would be similar to that of any large carnivorous lizard. So George duly arrived, was petted, admired and placed reverently in a large cage which had been prepared for him, with a special pond of his own. This amenity he appeared to appreciate fully, for the moment he was released into his quarters he made straight for the pond and plunged in. He spent half an hour or so squatting in the water, occasionally ducking his head beneath the surface for a few minutes at a time and peering thoughtfully about the bottom of the pond. That evening we gave him a dead rat, which he regarded with considerable loathing. Then we tried him on a young chicken, with the same result. Fish he retreated from as if it were some deadly poison, and we were in despair, for we could think of nothing else that he might like. Just when we were convinced that George was going to starve himself to death, John had an idea. He went off and fetched a handful of fat garden snails and tossed them into George's pond. George, who had been sitting on a tree trunk at the back of the cage and looking very regal, eyed this floating, frothing largesse with his head on one side. Then he came down to the pool, slid into the water and nosed interestedly at a snail, while we watched hopefully. Delicately he picked up the snail in his mouth and, throwing back his head, allowed it to slide to the back of his mouth. Now that his mouth was open I could see that he had the most astonishing teeth I had ever seen in a lizard: teeth that were, of course, perfectly adapted for eating snails. Those in the front of the mouth were fairly small, pointed and inclined slightly backwards into the mouth. These were the grasping teeth, as it were. Once they had hold of the snail, the lizard threw back his head so that the mollusc slid and rested on the teeth at the back of the mouth. These were huge, shoe-box-shaped molars with carunculated surfaces which looked more like miniature elephant's teeth than anything else. With the aid of his tongue, George manoeuvred the snail until it rested between these ponderous molars, and then closed his jaws slowly. The snail cracked and splintered, and when he was quite sure that the shell was broken he shifted the whole into the centre of his mouth and, by careful manipulation of his tongue, extracted all the bits of broken shell and spat them out. Then the smooth, shell-less body of the snail was swallowed with every evidence of satisfaction. The complete process took about a minute and a half, after which George sat there for a bit, licking his lips with his black tongue, and musing to himself.

After a time he leant forward and daintily picked up another snail, which he despatched in the same manner. Within half an hour he had eaten twelve of these molluscs, and we were jubilant, for, having now discovered George's preference, we knew there would be no more difficulty in keeping him.

It is always a relief when a reptile starts to feed itself, for if it refuses food for a certain length of time it has to be force-fed, and that is a tricky and unpleasant job. Many of the constricting snakes refuse food on their arrival, and have to be force-fed until they have settled down, but it is not an operation one relishes, since, with their fragile jaws and teeth, it is very easy to break something and thus set up an infection in the mouth. I think the worst force-feeding job we ever had was with a pair of young gharials. These are Asiatic members of the crocodile family, and in the wild state feed on fish. Instead of the strong, rather blunt jaws of the alligators and crocodiles, the gharial's jaws are long and very slender, resembling a beak more than anything. Both the jaws and the teeth are very fragile, the teeth especially so, for they appear to fall out if you look at them. In consequence, when our two young gharials arrived and steadfastly refused all food, including live fish in their pond, our hearts sank, as we realized we would have to force-feed them. The process was tedious, protracted and difficult, and had to be done once a week for a year before the gharials would feed on their own. First, you take a firm grip on the back of the creature's neck and his tail. Then you lift him out of the tank and place him on a convenient flat surface. Whoever is helping you, then slides a flat, smooth piece of wood between the jaws at the back of the mouth, immediately behind the last teeth. When the jaws are prised a little apart, you slightly release your grip on the reptile's neck and slide your hand forward, push your thumb and forefinger between the jaws and hold them apart. This is generally much easier than it sounds. The other person then arms himself with a long, slender stick and a plateful of raw meat chunks or raw fish. Impaling a piece of meat or fish on the end of the stick, he inserts it into the reptile's mouth and pushes it towards the back of the throat. This is the tricky part, for in all members of the crocodile family the throat is closed by a flap of skin: this arrangement allows the creature to open its mouth beneath the surface without swallowing vast quantities of water. The food has to be pushed past this flap of skin and well down into the throat. Then you massage the throat until you feel the food slide down into the stomach. As I say, it is a tedious task, as much for the gharial as for you.

By and large, the creatures that seem to cause the least trouble in the Reptile House are the amphibians. They usually feed well, and they do not seem to suffer from the awful variety of cankers, sores and parasites that snakes and lizards contract, though I must admit they can come up with one or two choice complaints of their own on occasions, just to enliven things for you. The pipa toads were a good example of this. These extraordinary creatures come from British Guiana, and look, quite frankly, like nothing on earth. Their bodies are almost rectangular, with a leg at each corner, so to speak, and a pointed bit between the front legs that indicates where the head is supposed to be. The whole affair is very flattened and a dark blackish brown colour, so the creature looks as though it had met with a nasty accident some considerable time ago and has been gently decomposing ever since. The most extraordinary thing about these weird beasts is their breeding habits, for the pipa toads carry their young in pockets. During the breeding season the skin on the female's back becomes thickened, soft and spongy, and then she is ready for mating. The male clasps her, and as soon as she is ready to lay she protrudes a long ovipositor which curves up on to her back, underneath the male's stomach. As the eggs appear, he fertilizes them and presses them into the spongy skin of the back. They sink in until only a small proportion of the egg is above the surface of the skin. This exposed portion of egg hardens. So, inside their individual pockets, the tadpoles undergo their entire metamorphosis until they change into tiny replicas of their parents. When they are ready to hatch, the hardened top of the shell comes loose, and the tiny toads push it back and climb out, looking rather like someone getting out of a bubble-car.

I had once been fortunate enough to witness the hatching of some baby pipa toads, and I was anxious to see if we could breed them in the Zoo. So I ordered a pair from a dealer, and on their arrival duly installed them in the Reptile House. We kept them in a large aquarium full of water, for, unlike other toads, pipas are entirely aquatic. They settled down very well, and were soon devouring monstrous great earthworms by the score. All we had to do now, I thought, was to wait for them to breed. One morning John came to me and said that one of the pipas had apparently bruised itself on the stomach, though he could not see how this had happened. I examined the toad and discovered that what appeared to be a bruise was something which looked like a gigantic blood blister. It was difficult to know what to do. If the toads had not been acquatic and had had dry skin, I would have anointed the area with penicillin. Within twenty-four hours both pipas were dead, their bodies covered with the red blisters which were full of blood and mucous. I sent them away for a post-mortem, and the report came back that they were suffering from an obscure disease called red-leg. I had a strong feeling that this had something to do with the water in which they had been kept: it was ordinary tap water but rather acid. So I purchased another pair of pipas, and this time we kept them in pond water only. This has, so far, proved successful, and, at the time of writing, both toads are flourishing. With a bit of luck, I might get around to breeding pipa toads yet, unless they can think up something new to frustrate me.

Another amphibian with almost as fascinating breeding habits as the pipas is the little pouched frog. We had five of these delightful tubby little frogs, handsomely marked in green and black, which were brought to us from Ecuador. They did very well, eating prodigiously, but they showed no signs of wanting to breed. So we moved them into a bigger tank where they had more land and water space, and this did the trick. Out of the breeding season, the female's pouch, which is on her lower back, is scarcely noticeable. If you look closely, you can see a faint line down the skin, with a slightly puckered edge, as if at one time the skin had been torn and healed up rather badly. However, when the breeding season comes round, the slit becomes much more obvious. The frogs begin to sing to each other, and presently you will

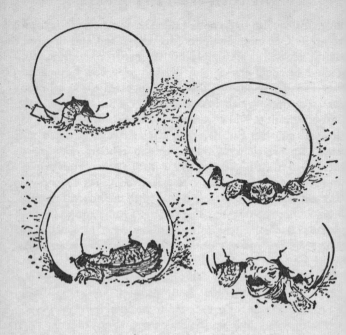

see the females going off into quiet corners and indulging in a very curious action. They manage, by great contortion, to get one hind leg at a time up over their backs, insert the toes into their pouches and proceed to stretch the skin. When the pouch is stretched to their satisfaction, they are ready to breed. The method by which they put the eggs into the pouch is still a mystery to me, for, unfortunately, I missed the actual egg-laying. The next thing we knew was that the female had a bulging pouchful of spawn which protruded from her back and made her look as though she had been disembowelled. The female carries the eggs around until she knows, by some means or other, that the tadpoles are ready to hatch, whereupon she goes and sits in the water. The tadpoles wriggle free of the gelatine-like spawn and swim off on their own, the mother taking no further interest in them. We found that the tadpoles did very well on strips of raw meat and white worms, the tiny worms that fish fanciers breed as food. When they grew their legs and came out on land, we fed them on fruit flies and tiny earthworms, until they were old enough to graduate to house-flies and bluebottles.

Amphibians are much easier to breed than reptiles, for you do not have to worry about the moisture. Reptiles lay eggs with a parchment-like shell which is either soft or hard. If the temperature of the cage is not right, and if the moisture content of the air is too great or not enough, the contents of the egg will either dry up or else go mildewed. Although we have had some successes with hatching reptile eggs, the chances against are always ninety to one. One success we did achieve, of which we were rather proud, was in hatching some Greek tortoise eggs. The Greek tortoise is probably one of

the commonest pets, and they invariably lay eggs with monotonous regularity, but these very seldom hatch. Thinking that this batch of eggs was going to be no more successful than all the others had been, John did not worry overmuch about them. He buried them in the sand at the bottom of one of the cages which had what he thought was a suitable temperature. Week after week passed, and eventually he forgot all about them. He was, therefore, considerably astonished one morning to find a baby tortoise perambulating about the cage. He called me and we dug up the rest of the eggs. Out of the six, four were in the process of hatching. In one egg the baby was almost out, but in the other three the babies had only just started to breach the shell. We placed them in a small aquarium on a saucer of sand, in order to watch the hatching more conveniently. The eggs were almost the size of ping-pong balls, and much the same shape; the parchment-like shell was tough, and it was clearly an exhausting job for the babies to break out of their prisons. The one who had made the biggest hole in his shell could be seen quite plainly inside, as he twisted round and round, now using his front feet and now his back ones to enlarge the hole. On his nose he had the little horny 'beak' which baby tortoises are supposed to use to make the first breach in the shell; this later drops off. But I did not see this one use his 'beak' at all – all the hard work was done with the front and hind legs, with frequent pauses for him to regain his strength. It took him three-quarters of an hour to break out, and then the egg split in half and he trundled off across the sand, wearing one half on top of his carapace, like a hat. When they emerged from the egg, their shells were spongy, misshapen and extremely soft, and they were each the diameter of a two-shilling piece. However, after an hour or so a change had taken place; it was as though someone had inflated them with a bicycle pump. The shell had filled out, and, instead of being flattened, it was now handsomely domed and looked much harder, although it was, in actual fact, still as soft as damp cardboard. They were now so much larger than the egg that, unless I had watched them hatching myself, I would have said they could not possibly have emerged so recently from such a small prison. I noticed that their nails, when they hatched, were very long and sharp, presumably to help them break through the egg shell. Within a very short time, though, they had worn down to a normal length.

I had spent several hours watching this hatching process, and it was worth every minute of it. I had the greatest admiration for these rotund and earnest little tortoises, for breaking out of the egg was no easy matter. What amused me most, I think, was the way – after he had been using the hind feet to enlarge the hole – the tiny reptile would swivel round inside the shell, and the next moment a minute, wrinkled and rather sad little face would be poked through the hole in the shell, as if the tortoise wanted to reassure himself that the outside world was still there and still as attractive as it had been when he last looked. We were very lucky to have been able to hatch these tortoises, but what was even luckier was the fact that Ralph Thompson, who illustrated this book, happened to be staying with me at the time, and was thus able to draw the whole of the hatching process from start to finish, which he assured me he thoroughly enjoyed, in spite of the fact that, owing to the high temperature in the Reptile House, his glasses kept misting over.

CHAPTER FOUR

CLAUDIUS AMONG THE CLOCHES

Dear Mr Durrell,
Do you ever stuff your animals? If you ever wanted to stuff your
animals I could stuff them for you, as I have a great experience in
stuffing animals . . .

ON acquiring new animals, one of the many problems that face you is the process of settling them in, for until they have learnt to look upon their new cage as home, and have also learnt to trust you, they are unsettled. There are many different ways of making animals feel at home, and these vary according to the species. Sometimes special titbits have to be given, so that the animal forgets its fear of you in its eagerness for the food. With highly nervous creatures you may have to provide them with a box in which they can hide, or cover the front of the cage with sacking until they have decided that you mean them no harm. There are times when the most extraordinary methods have to be used to give an animal confidence and the trouble we had with Topsy was a case in point.

I was in an animal dealer's shop in the North of England one cold winter's day, looking around to see if he had anything interesting I could buy for the Zoo. As I walked round the shop I suddenly noticed a very dank, dark cage in one corner, and peering at me from behind the bars was one of the most pathetic little faces I had ever seen. It was coal-black with large, lustrous eyes that seemed to be perpetually full of tears. The fur surrounding this face was reddish-brown, short and thick like the pile on an expensive carpet. I looked closer and saw that the face belonged to a baby woolly monkey, one of the most charming of the South American primates. This one could not have been more than a few weeks old, and was far too young to have been separated from its mother. It crouched miserably on the floor of the cage, shivering and coughing, its nose streaming, its fur matted and tangled with filth. From the condition and smell of the cage I could see that it had enteritis as well as a cold which looked as though it was bordering on pneumonia. It was not an animal that anyone in their right senses would contemplate buying. But then it peered up at me with its great, dark eyes filled with despair, and I was lost. I asked the dealer how much he wanted for the baby. He said that he would not dream of selling it to me, as I was a good customer and the baby was sure to die. I replied that I realized the animal was a bad risk, but that if he would let me have it I would pay him if it lived, but not if it died. Rather reluctantly he agreed to this, and we bundled the plaintively squealing baby into a box full of straw, and I hurried back to Jersey with it. I knew that unless it was treated rapidly, it would die, and already it might well be too late.

On my return to Jersey, we put the baby, which someone christened Topsy, into a warm cage and examined her. First, I realized she would have to have antibiotic and vitamin injections to combat the enteritis and the cold. Secondly, her thick fur, matted with her own excreta, would have to be cleaned, for if it was left in that state she could develop skin rash and eventually lose all her fur. Our chief problem, though, was how to get Topsy to allow us to do these things. Most baby monkeys will, within a matter of hours, take to a human foster-parent, and they are generally no trouble at all. As Topsy's experience of human beings had obviously been of the worst possible kind, she threw herself in fits of screaming hysterics (as only a woolly monkey can) if we so much as opened the door of her cage. To manhandle her was, therefore, going to do more harm than good, and yet she had to have treatment or die. Then we had a brainwave: if Topsy would not accept us as foster parents, would she accept something else? How about a teddy-bear? We were all a bit doubtful about this, but we had to try something, and so we obtained one. The bear had a pleasant, if slightly vacuous expression, and was just about the size that Topsy's mother would be, so we put it in the cage and awaited results. At first, Topsy would not go near it, but at last her curiosity got the better of her and she touched it. As soon as she discovered that it was cuddly and furry, she took to it, and within half an hour was clinging to it with a fierce, possessive passion that was quite touching.

Now, a complete change came over Topsy. As long as she was clinging to her teddy-bear with arms, legs and tail, she lost her fear of human beings. We just simply lifted the bear out of the cage with Topsy stuck to it, like a limpet, and she would allow us to do what we liked. We were thus able to inject her and clean up her matted fur, and within a few days she was well on the road to recovery, and looked a different monkey. But then came another problem: as the days passed, the teddy-bear became more and more unhygienic, until finally we decided he would have to be removed from Topsy's cage to be washed and disinfected. So, to Topsy's extreme annoyance, we removed the bear. Immediately she threw a screaming fit. Of all the monkey family, the woolly monkeys have the most powerful and excruciating scream you have ever heard, a scream that goes through you and makes your blood run cold, like the screech of a knife on a plate, magnified a million times. We blocked our ears, and consoled ourselves with the thought that she would stop in about ten minutes when she realized that she was not going to have her bear back, but Topsy did not stop. She screamed solidly all morning, and by lunchtime our nerves were in shreds. There was

only one thing to do: we took the van and rushed down into the town and, after visiting several toy shops, managed to buy a teddy-bear closely resembling Topsy's original one. Then we hurried back to the Zoo and stuffed it hastily into Topsy's cage. She stopped in mid-scream, gave a loud squeak of joy and flung herself on to the new teddy-bear, wrapped her arms, legs and tail tightly round it, and immediately fell into a deep and exhausted sleep. After that, the teddy-bears took it in turn: while one was being washed, the other took over the duties of foster mother, and this arrangement Topsy found eminently satisfactory.

At last Topsy had grown so big that she was bigger than her teddy-bears, and we decided that we would have to wean her off them, as it were, for eventually she would have to go into a big cage with other woolly monkeys, and she could not take her bears with her. It was time, we felt, that she grew used to the idea of having a companion in the cage with her, and so we chose a large ginger guinea-pig of placid disposition and no brain. He was introduced into Topsy's cage, and at first she ignored him, except when he went too near to her precious bear, whereupon she would clout him. It was not long, however, before Topsy discovered that the guinea-pig had one great advantage over the bear as a sleeping companion, and that was it had built-in central heating. The guinea-pig – whom we now called Harold for convenient reference – took, I think, a rather dim view of all this. To begin with, if he possessed a thought in his head at all, that thought was food. Harold's life-work was to test the edibility of everything into which he came in contact, and he did not like having his life's work mucked about by a domineering woolly monkey. Topsy, on the other hand, had very strict ideas about the correct time to get up, go to bed, play, and so on, and she did not see why she should have to change these to fit in with Harold's feeding habits. It seemed to Harold that no sooner had he found a respectable piece of carrot, or something, than Topsy would decide it was bedtime, and he was seized by the hind leg and hauled off to their box of straw, in the most undignified manner. Here, to add insult to injury, Topsy would climb on to his back, wrap her arms, legs and tail tightly round him to prevent his escape, and sink into a deep sleep, looking like an outsize jockey on a small and very rotund ginger horse.

Another thing that Harold found rather disquieting was Topsy's firm conviction that, if given the opportunity, he would be able to leap about in the branches with the same agility that she herself displayed. She was sure that, if only she could get him *up* into the branches, he would turn out to be a splendid climber, but the job was to lift Harold off the ground. She could only spare one hand to hold him with, and he was fat, heavy and uncooperative. She would, after considerable effort, tuck him under one arm, and then start to climb, but before she was more than a few inches up the wire Harold would slip out from under her arm and plop back to the floor of the cage. Poor Harold, I think he suffered a great deal at Topsy's hands, but he served our purpose, for very soon Topsy had forgotten all about her teddy-bears, and was able to take her place in the big cage with the rest of the woolly monkeys. Harold was returned to the guinea-pig pen where he spends all day up to his knees in vegetables, champing his way through them with grim determination.

Another creature that gave us a certain amount of trouble during his settling-in period was Fred, a patas monkey from West Africa. He was a fully adult male, one of the largest patas I have ever seen, and he had been the personal pet of some people in England. How they managed to keep him up to that size without being severely bitten was a mystery, for Fred's canines were a good two inches long and as sharp as razors. Apparently, right up to the time that Fred came to us, he used to go into the house each evening and watch television.

But the really awful thing about Fred was his clothing. Patas monkeys are covered with thick, bright ginger-coloured fur, and Fred arrived wearing a knitted jumper in a startling shade of pillar-box red. This combination of colours made even the most un-sartorial members of the staff blench. The trouble was that Fred missed his television and his rides in the car, and decided that we were in some way responsible for depriving him of these, and so he loathed us all from the very start with complete impartiality. If anyone went near his cage he would leap at the wire and shake it vigorously, baring all his teeth in a ferocious grimace. Until, if ever, he showed any signs of trusting and liking us, we could do nothing about removing his terrible jacket. Fred just sat among the branches in his cage, wearing his scarlet jacket and showing no signs of forgiving us. The trouble was that, as the days passed, the jumper grew more and more grubby and dishevelled until he looked as though he had just emerged from some poverty-stricken slum. We tried every method to rid him of this insanitary garment, but without success. Fred seemed rather proud of it, and would become very annoyed if we tried to take it off him. We began to wonder how long it would take the wool to disintegrate naturally and fall off, but whoever had knitted the jumper had chosen really tough wool, and it was obvious that it would be several years before it fell to pieces. Then fate played into our hands. We had a heat-wave, and the temperature in the Mammal House where Fred lived soared up. At first, he enjoyed it, but soon it became too much even for him, and we noticed that he was pulling meditatively at his jumper. The next morning we found the offending garment hanging neatly over a branch in Fred's cage, and managed to hook it out with the aid of a long stick. From that day onwards, Fred grew increasingly placid; he will never be really trustworthy, but at least he is now less inclined to treat human beings as his enemies.

Still another creature that gave us a certain amount of trouble in the early stages was Millicent, the Malabar squirrel. Malabars are the largest members of the squirrel family, and hail from India. They measure about two feet in length, with

sturdy bodies and long, very bushy tails. Their undersides are saffron yellow, their upper parts a rich mahogany red, and they have very large ear-tufts that are like a couple of black sporrans perched on their heads. They are, like all squirrels, very alert, quick moving and inquisitive, but, unlike most squirrels, they do not have that nervous desire to gnaw everything with which they come into contact. The exception to this was, of course, Millicent. Her view was that nature had provided her with a pair of very prominent, bright orange teeth for the sole purpose of demolishing any cage in which she was confined. This was not from any desire to escape, because having gnawed a large hole in one side of the cage she would then move over to the other side and start all over again. She cost us a small fortune in repairs until we had a cage specially lined with sheet metal, and thus put a stop to her activities. However, feeling that she would miss her occupational therapy, we gave her large logs of wood, and she proceeded to gnaw her way through these, like a buzz saw.

At first, Millicent was anything but tame, and would not hesitate to bury her teeth in your finger, should you be foolish enough to give her the chance. No amount of bribery on our part, with the aid of such things as mushrooms and acorns, would make her any the less savage, and we came to the conclusion that she was just one of those animals which never become tame. But then a peculiar thing happened: Millicent was found one day lying in the bottom of her cage in a state of collapse. She had no obvious symptoms, and it was a little difficult to tell exactly what was wrong with her. When I find an animal suffering from some mysterious complaint like this, I do two things: I give it an antibiotic and keep it very warm. So Millicent had an injection and was moved down to the Reptile House, for this is the only place where the heat is kept on throughout the summer months.

Within a few days Millicent was recovering satisfactorily, but was still languid. The extraordinary fact was the change in her character. From being acutely anti-human, she had suddenly become so pro-*Homo sapiens* that it was almost embarrassing. You had only to open her cage door and she would rush out into your arms, nibbling your fingers gently and peering earnestly into your face, her long whiskers quivering with emotion. She liked nothing better than to lie along your arm, as though it were the branch of a tree, and in this position doze for hours if you let her. Since she was now such a reformed character, she was allowed out of her cage first thing each morning, to potter round the Reptile House. Millicent very soon discovered that the tortoise pen provided her with everything a self-respecting Malabar could want: there was an infra-red lamp that cast a pleasant, concentrated heat; there were the backs of the giant tortoises which made ideal perches; and there was an abundance of fruit and vegetables. So the giant tortoises would move ponderously round their pen, while Millicent perched on their shells. Occasionally, when one of them found a succulent piece of fruit and was just stretching out his neck to engulf it, she would hop down from his back, pick up the fruit, and jump back on to the shell

again before the tortoise really knew what was happening. When the time came that Millicent was well enough to return to the Small Mammal House, I think the giant tortoises were glad to see the back of her, for not only had she been an additional weight on their shells, but the constant disappearance of titbits from under their very noses was having a distressing effect on their nerves.

It is amazing how wild-caught animals (as opposed to hand-reared ones) differ in settling down in captivity. Some take a considerable time to adjust themselves, while others, from the moment of arrival, carry on as if they had been born in the Zoo. A dealer sent us a pair of brown woolly monkeys which he had just received direct from Brazil. We found that the male was a magnificent specimen, fully adult, and must have been about twelve or fourteen years old. We were not very pleased with this, for an adult monkey of that age would, we felt, take a long time to adjust itself to captivity, and might even pine and die. We released him into his cage with his mate, and went and fetched them some fruit and milk. As soon as he saw these, he became very excited, and when the door of the cage was opened, to our complete astonishment, he came straight down and ate and drank while we were still holding the dishes, as if he had been with us for years instead of a matter of minutes. Right from the start he was perfectly tame, and ate well and seemed to thoroughly enjoy his new life.

There are many creatures which, on being settled in, make the most determined attempts to escape from their cages, not because they want their freedom but simply because they miss their old territory: the travelling crate to which they have grown used and which they look upon as their home. I have known an animal that was removed from its tiny, travelling crate and placed in a spacious, well-appointed cage spend three days endeavouring to break out, and when it was finally successful it made a bee-line to its old travelling box and was found sitting inside it. The only answer to this problem was to place the travelling crate inside the new cage. This we did, and the animal used it thereafter as its bedroom and settled down quite happily.

There are again some creatures, of course, which, when they manage to escape, present you with considerable problems. For instance, there was the night I shall never forget, when Claudius, the South American tapir, contrived to find a way out of his paddock. The person who had been in to give him his night feed had padlocked the gate carefully but without sliding the bolt into position. Claudius, having a nocturnal perambulation round his territory, found to his delight that the gate which he had hitherto presumed to be invulnerable now responded to his gentle nosings. He decided that this was a very suitable night to have a short incursion into the neighbouring countryside. It was a suitable night from Claudius's point of view, because the skies were as black as pitch and the rain was streaming down in torrents that I have rarely seen equalled outside the tropics.

It was about quarter past eleven, and we were all on the point of going to bed, when a rather harassed and extremely wet motorist appeared and beat upon the front door. Above the roar of the rain, he said that he had just seen a big animal in the headlights of his car, which he felt sure must be one of ours. I asked him what it looked like, and he said it looked to him like a misshapen Shetland pony with an elephant's trunk. My heart sank, for I knew just how far and how fast Claudius could gallop if given half a chance. I was in my shirt-sleeves and only wearing slippers, but there was no time to change into more suitable attire against the weather, for the motorist had spotted Claudius in a field adjoining our property and I wanted to catch up with him before he ventured too far afield. I rushed round to the cottage and harried all those members of the staff who lived in. In various stages of night attire they all tumbled out into the rain and we headed for the field into which the motorist assured us our tapir had disappeared. This was quite a large field and belonged to our nearest neighbour, Leonard du Feu. Leonard had proved himself to be the most long suffering and sympathetic of neighbours, and so I was determined that Claudius was not going to do any damage to his property if we could possibly avoid it. Having made this mental resolve, I then remembered to my horror that the field in which Claudius was reputedly lurking had just recently been carefully planted out by Leonard with anemones. I could imagine what Claudius's four hundredweight could do to those carefully planted rows of delicate plants, particularly as, owing to his short-sightedness, his sense of direction was never too good at the best of times.

We reached the field, soaked to the skin, and surrounded it. There, sure enough, stood Claudius obviously having the best evening out he had had in years. The wet as far as he was concerned was ideal: there was nothing quite like a heavy downpour of rain to make life worth while. He was standing there, looking like a debauched Roman emperor under a shower, meditatively masticating a large bunch of anemones. When he saw us, he uttered his greeting – a ridiculous, high-pitched squeak similar to the noise of a wet finger being rubbed over a balloon. It was quite plain that he was delighted to see us and hoped that we would join him in his nocturnal ramble, but none of us was feeling in any mood to do this. We were drenched to the skin and freezing cold, and our one ambition was to get Claudius back into his paddock with as little trouble as possible. Uttering a despairing and rather futile cry of 'don't tread on the plants', I marshalled my band of tapir-catchers and we converged on Claudius in a grim-faced body. Claudius took one look at us and decided from our manner and bearing that we did not see eye to eye with him on the subject of gambolling about in other people's fields at half past eleven on a wet night, and so he felt that, albeit reluctantly, he would have to leave us. Pausing only to snatch another mouthful of anemones, he set off across the field at a sharp gallop, leaving a trail of destruction behind him that could only have been emulated by a runaway bulldozer. In our slippered feet, clotted with mud, we stumbled after him. Our speed was reduced not only by the mud but by the fact that we were trying to run between the rows of flowers instead of on them. I remember making a mental note as I ran that I would ask Leonard in future to plant his rows of flowers wider apart, as this would facilitate the recapture of any animal that escaped. The damage Claudius had done to the flowers was bad enough, but worse was to follow. He suddenly swerved, and instead of running into the next field, as we had hoped (for it was a grazing meadow), he ran straight into Leonard du Feu's back-garden. We pulled up short and stood panting, the rain trickling off us in torrents.

'For God's sake,' I said to everyone in general, 'get that bloody animal out of that garden before he wrecks it.' The words were hardly out of my mouth when from inside the garden came a series of tinkling crashes which told us too clearly that Claudius, trotting along in his normal myopic fashion, had ploughed his way through all Leonard's cloches. Before we could do anything sensible, Claudius having decided that Leonard's garden was not to his liking, crashed his way through a hedge, leaving a gaping hole in what hitherto had been a nice piece of topiary, and set off into the night at a brisk trot. The direction he was taking presented yet another danger, for he was heading straight for our small lake. Tapirs in the wild state are very fond of water and they are excellent swimmers and can submerge themselves for a considerable length of time. The thought of having to search for a tapir in a quarter of an acre of dark water on a pitch black, rainy night made the thought of hunting for a needle in a haystack pale into insignificance. This thought struck the others members of my band at the same moment, and we ran as we had never run before and just succeeded at the very last minute in heading off Claudius. Coming up close to his rotund behind, I launched myself in a flying tackle and, more by luck than judgement, managed to grab him by one hind leg. In thirty seconds I was wishing that I had not. Claudius kicked out and caught me a glancing blow on the side of the head, which made me see stars, and then revved up to a gallop,

dragging me ignominiously through the mud, but by now I was so wet, so cold, so muddy and so angry that I clung on with the determination of a limpet in a storm. My tenacity was rewarded, for my dragging weight slowed Claudius down sufficiently to allow the others to catch up, and they hurled themselves on various portions of his anatomy. The chief difficulty with a tapir is that there is practically nothing on which to hold: the ears are small and provide a precarious grip, the tail is minute, there is no mane, so really the only part you can grip with any degree of success are its legs, and Claudius' legs were fat and slippery with rain. However, we all clung on grimly, while he bucked and kicked and snorted indignantly. As one person loosened his hold, another one would grab on until eventually Claudius decided he was using the wrong method of discouragement. He stopped pirouetting about, thought to himself for a moment and then just simply lay down and looked at us.

We stood round him in a sodden, exhausted circle and looked at each other. There were five of us and four hundred-weight of reluctant tapir. It was beyond our powers to carry him, and yet it was quite obvious that Claudius had no intention of helping us in any way. He lay there with a mulish expression on his face. If we wanted to get him back to the Zoo, it implied, we would jolly well have to carry him. We had no more reinforcements to call on, and so it appeared that we had reached an impasse. However, as Claudius was prepared to be stubborn, I was prepared to be equally so. I sent one of my dripping team back to the Zoo for a rope. I should, of course, have brought this very necessary adjunct of capture with me, but in my innocence I had assumed that Claudius could be chivied back to his paddock with no more trouble than a domestic goat. When the rope arrived, we attached it firmly round Claudius' neck, making sure that it was not a slip-knot. I thought one drenched member of the staff was heard to mutter that a slip-knot would be ideal. Then two of us took hold of the rope, two more took hold of his ears, and the fifth took hold of his hind legs, and by the application of considerable exertion we raised him to his feet and wheel-barrowed him all of ten feet, before he collapsed again. We had a short pause to regain our breath and started off again. Once more we carted him for about ten feet, in the process of which I lost a slipper and had my hand heavily trodden on by one of the larger and weightier members of my team. We rested again, sitting dejectedly and panting in the rain, longing for a cigarette and unanimously deciding that tapirs were animals that should never in any circumstances have been invented.

The field in which these operations took place was large and muddy. At that hour of night, under the stinging rain, it resembled an ancient tank-training ground which had been abandoned because the tanks could no longer get through it. The mud in it appeared to have a glue-like quality not found elsewhere in the Island of Jersey. It took us an hour and a half to get Claudius out of that field, and at the end of it we felt rather like those people must have felt who erected Stonehenge – that none of us was ruptured was a miracle. A final colossal effort and we hauled Claudius out of the field and over the boundary into the Zoo. Here we were going to pause for further recuperation, but Claudius decided that since we had brought him back into the Zoo grounds and would, it appeared, inevitably return him to his paddock, it would be silly to delay. He suddenly rose to his feet and took off like a rocket, all of us desperately clinging to various parts of his body. It seemed ludicrous that for an hour and a half we should have been making the valiant attempt to get him to move at all and now we were clinging to his fat body in an effort to slow him down for fear that in his normal blundering way he would run full tilt into one of the granite archways and hurt or perhaps even kill himself. We clung to him like sucker-fish to a speeding shark, and, to our intense relief, managed to steer our irritating vehicle back into its paddock, without any further mishap, and so we returned to our respective bedrooms, bruised, cold and covered with mud. I had a hot bath to recuperate, but as I lay in it drowsily, I reflected that the worst was yet to come: the following morning I had to telephone Leonard du Feu and try to apologize for half an acre of trampled anemones and twelve broken cloches.

Jacquie, as always, was unsympathetic. As I lay supine in the comforting warmth of the bath, she placed a large whisky within easy reach and summed up the night's endeavour:

'It's your own fault,' she said, 'you would get this blasted Zoo.'

CHAPTER FIVE

THE NIGHTINGALE TOUCH

Dear Mr Durrell,
You are the most evil man I know. All God's creatures should have their freedom, and for you to lock them up is against His Will. Are you a man or a devil? You would be locked up in prison for the rest of your life if I had my way . . .

WHETHER you run a pig farm, a poultry farm, a mink farm or a zoo, it is inevitable that occasionally your animals will damage themselves, become diseased, and eventually that they will die. In the case of death, however, the pig, mink or poultry farmer is in a very different position from the person who owns a zoo. Someone who visits a pig farm and inquires where the white pig with the black ears has gone, is told that it has been sent to market. The inquirer accepts this explanation without demur, as a sort of porcine kismet. This same person will go to a zoo, become attracted to some creature, visit it off and on for some time, and then, one day, will come and find it missing. On being told that it has died, they are immediately filled with the gravest suspicion. Was it being looked after properly? Was it having enough to eat? Was the vet called in? And so on. They continue in this vein, rather like a Scotland Yard official questioning a murder suspect. The more attractive the animal, of course, the more searching do their inquiries become. They seem to be under the impression that, while pigs, poultry or mink die or are killed as a matter of course, wild creatures should be endowed with a sort of perpetual life, and only some gross inefficiency on your part has removed them to a happier hunting ground. This makes life very difficult, for every zoo, no matter how well-fed and cared-for are its animals, has its dismal list of casualties.

In dealing with the diseases of wild animals you are venturing into a realm about which few people know anything, even qualified veterinary surgeons, so a lot of the time one is working, if not in the dark, in the twilight. Sometimes the creature contracts the disease in the zoo, and at other times it arrives with the disease already well established, and it may well be a particularly unpleasant tropical complaint. The case of Louie, our gibbon, was a typical one.

Louie was a large, black gibbon with white hands, and she had been sent to us by a friend in Singapore. She had been the star attraction in a small R.A.F. zoo where – to judge by her dislike of humans, and men in particular – she had received some pretty rough handling. We put her in a spacious cage in the Mammal House, and hoped that, by kind treatment, we would eventually gain her confidence. For a month all went well. Louie ate prodigiously, actually allowing us to stroke her hand through the wire, and would wake us every morning with her joyous war cries, a series of ringing 'whoops' rising to a rapid crescendo and then tailing off into what sounded like a maniacal giggle. One morning, Jeremy came to me and said that Louie was not well. We went down to have a look at her, and found her hunched up in the corner of her cage, looking thoroughly miserable, her long arms wrapped protectively round her body. She gazed at me with the most woebegone expression, while I racked my brains to try to discover what was wrong with her. There seemed to be no signs of a cold, and her motions were normal, though I noticed her urine was very strong and had an unpleasant pungent smell. This indicated some internal disorder, and I decided to give her an antibiotic. We always use Tetramycin, for this is made up in a thick, sweet, bright red mixture, which we have found by experience that very few animals can resist.

Some monkeys would, if allowed, drink it by the gallon. At first, Louie was clearly so poorly that she would not even come to try the medicine. At last, after considerable effort, we managed to attract her to the wire, and I tipped a teaspoonful of the mixture over one of her hands. Hands, of course, are of tremendous importance to such an agile, arboreal creature as a gibbon, and Louie was always very particular about keeping her hands clean. So to have a sticky pink substance poured over her fur was more than she could endure, and she set to work and licked it off, pausing after each lick to savour the taste. After she had cleaned up her hand to her satisfaction,

I pushed another teaspoonful of Tetramycin through the wire, and to my delight she drank it greedily. I continued this treatment for three days, but it appeared to be having no effect whatsoever, for Louie would not eat and grew progressively weaker. On the fourth day I caught a glimpse of the inside of her mouth, and saw that it was bright yellow. It seemed obvious that she had jaundice, and I was most surprised, for I did not know that apes or monkeys could contract this disease. On the fifth day Louie died quite quietly, and I sent her pathetic corpse away to have a post-mortem done, to make sure my diagnosis was correct. The result of the post-mortem was most interesting. Louie had indeed died of jaundice, but this had been caused by the fact that her liver was terribly diseased by an infestation of filaria, a very unpleasant tropical sickness that can cause, among other things, blindness and elephantiasis. We realized, therefore, that, whatever we had tried to do, Louie had been doomed from the moment she arrived. It was typical that Louie, on arrival, had displayed no symptoms of disease, and had, indeed, appeared to be in quite good condition.

This is one of the great drawbacks of trying to doctor wild animals. A great many creatures cuddle their illnesses to themselves, as it were, and show no signs of anything being wrong until it is too late – or almost too late – to do anything effective. I have seen a small bird eat heavily just after dawn, sing lustily throughout the morning, and at three o'clock in the afternoon be dead, without having given the slightest sign that anything was amiss. Some animals, even when suffering from the most frightful internal complaints, look perfectly healthy, eat well, and display high spirits that delude you into believing they are flourishing. Then, one morning, it looks off-colour for the first time, and, before you can do anything sensible, it is dead. And, of course, even when a creature is showing obvious symptoms of illness, you have to make up your mind as to the cause. A glance at any veterinary dictionary will show a choice of several hundred diseases, all of which have to be treated in a different manner. It is all extremely frustrating.

Generally, you have to experiment to find a cure. Sometimes these experiments pay off in a spectacular way. Take the case of the creeping paralysis, a terrible complaint that attacks principally the New World monkeys. At one time there was no remedy for this, and the disease was a scourge that could wipe out your entire monkey collection. The first symptoms are very slight: the animal appears to have a certain stiffness in its hips. Within a few days, however, the creature shows a marked disinclination to climb about, and sits in one spot. At this stage both hind limbs have become paralysed, but still retain a certain feeling. Gradually the paralysis spreads until the whole of the body is affected. At one time, when the disease reached this stage, the only thing to do was to destroy the animal.

We had had several cases of this paralysis, and lost some beautiful and valuable monkeys as a result. I had tried everything I could think of to effect a cure. We massaged them, we changed the diet, we gave them vitamin injections, but all to no purpose. It worried me that I could not find a cure for this unpleasant disease, since watching a monkey slowly becoming more paralysed each day is not a pretty sight.

I happened to mention this to a veterinary surgeon friend, and said that I was convinced the cause of the disease was dietary, but that I had tried everything I could think of without result. After giving the matter some thought, my friend suggested that the monkeys might be suffering from a phosphorus deficiency in their diet, or rather that, although the phosphorus was present, their bodies were unable, for some reason, to assimilate it. Injections of D_3 were the answer to this, if it was the trouble. So the next monkey that displayed the first signs of the paralysis was hauled unceremoniously out of its cage (protesting loudly at the indignity) and given an injection of D_3. Then I watched it carefully for a week, and, to my delight, it showed distinct evidence of improvement. At the end of the week it was given another injection, and within a fortnight it was completely cured. I then turned my attention to a beautiful red West African patas monkey, who had had the paralysis for some considerable time. This poor creature had become completely immobile, so that we had to lift up her head when she fed. I decided that, if D_3 worked with her, it would prove beyond all doubt that this was the cure. I doubled the normal dose and injected the patas; three days later I gave her another massive dose. Within a week she could lift her head to eat, and within a month was completely cured. This was a really spectacular cure, and convinced me that D_3 was the answer to the paralysis. When a monkey now starts to shuffle, we no longer have that sinking feeling, knowing that it is the first step towards death; we simply inject them, and within a short time they are fit and well again.

Another injection that we use a lot with conspicuous success is vitamin B_{12}. This acts as a general pick-me-up and, more valuable still, as a stimulant to the appetite. If any animal looks a bit off-colour, or starts to lack interest in its food, a shot of B_{12} soon pulls it round. I had only used this product on mammals and birds, but never on reptiles. Reptiles are so differently constructed from birds and mammals that one has to be a bit circumspect in the remedies you employ for them, as what may suit a squirrel or a monkey might well kill a snake or a tortoise. However, there was in the Reptile House a young boa constrictor which we had obtained from a dealer some six months previously. From the day it arrived it had shown remarkable tameness, but what worried me was that it steadfastly declined to eat. So, once a week, we had to haul the boa out of its cage, force open its mouth and push dead rats or mice down its throat, a process which he did not care for, but which he accepted with his usual meekness. Force-feeding a snake like this is always a risky business, for, however carefully you do it, there is always the chance that you might damage the delicate membranes in the mouth, and thus set up an infection which would quickly turn to mouth canker, a disease to which snakes are very prone, and which is very difficult to cure. So, with a certain amount of trepidation, I decided to give the boa a shot of B_{12} and see what happened. I injected halfway down his body, in the thick muscular layer that covers the backbone. He did not appear to even notice it, lying quite quietly coiled round my hand. I put him back into his cage and left him. Later on that day he did not seem to be any worse for his experience, and I suggested to John that he put some food in the cage that night. John placed two rats inside, and in the morning reported to me delightedly that not only had the boa eaten the rats but had also actually struck at his hand when he had opened the cage. From that moment on, the boa never looked back. As it had obviously only done good to the snake, I experimented with B_{12} on other reptiles. Lizards and tortoises I found benefited greatly from periodical shots, especially in the colder weather, and on several occasions the reptiles concerned would certainly have died but for the injections.

Wild animals, of course, make the worst possible patients in the world. Any nurse who thinks her lot is a hard one, handling human beings, should try her hand at a bit of wild animal nursing. They are rarely grateful for your ministrations, but you do not expect that. What you do hope for (and never, or hardly ever, receive) is a little cooperation in the matter of taking medicines, keeping on bandages, and so forth. After the first few hundred bitter experiences you reconcile yourself to the fact that every administration of a medicine is a sort of all-in wrestling match, in which you are more likely to apply more of the healing balm to your own external anatomy than to the interior of your patient. You soon give up all hope of keeping a wound covered, for nothing short of encasing your patient entirely in plaster of Paris is going to prevent it from removing the dressings within thirty seconds of their application. Monkeys are, of course, some of the worst patients. To begin with, they have, as it were, four hands with which to fight you off, or remove bandages. They are very intelligent and highly strung, on the whole, and look upon any medical treatment as a form of refined torture, even when you know it is completely painless. Being highly strung means that they are apt to behave rather like hypochondriacs, and quite a simple and curable disease may kill them because they just work themselves into a state of acute melancholy and fade away. You have to develop a gay, hearty bedside manner (rather reminiscent of a Harley Street specialist) when dealing with a mournful monkey which thinks he is no longer for this world.

Among the apes, with their far superior intelligence, you are on less shaky ground, and can even expect some sort of cooperation occasionally. During the first two years of the Zoo's existence we had both the chimps, Chumley and Lulu, down with sickness. Both cases were different, and both were interesting.

One morning I was informed that Lulu's ear was sticking out at a peculiar angle, but that, apart from this, she looked all right. Now Lulu's ears stuck out at the best of times, so I felt it must be something out of the ordinary for it to be so noticeable. I went and had a look at her and found her squatting on the floor of the cage, munching an apple with every sign of appetite, while she gazed at the world, her sad, wrinkled face screwed up in intense concentration. She was carefully chewing the flesh of the apple, sucking at it noisily, and then, when it was quite devoid of juice, spitting it into her hand daintily, placing it on her knee and gazing at it with the air of an ancient scientist who has, when he is too old to appreciate it, discovered the elixir of life. I called to her and she came over to the wire, uttering little breathless grunts of greeting. Sure enough, her ear looked most peculiar, sticking out at right angles to her head. I tried to coax her to turn round so that I could see the back of the ear, but she was too intent in putting her fingers through the wire and trying to pull the buttons off my coat. There was nothing for it but to get her out, and this was a complicated procedure, for Chumley became most jealous if Lulu went out of the cage without him. However, I did not feel like having Chumley as my partner during a medical examination. So, after much bribery, I managed to lure him into their bedroom and lock him in, much to his vocal indignation. Then I went into the outer cage, where Lulu immediately came and sat on my lap and put her arms round me. She was an immensely affectionate ape, and had the most endearing character. I gave her a lump of sugar to keep her happy, and examined the ear. To my horror, I found that, behind the ear on the mastoid bone, there was an immense swelling, the size of half an orange, and the skin was discoloured a deep purplish black. The reason this had not been noticed in the early stages was that Lulu had thick hair on her head, and particularly behind her ears, so that – until the swelling became so large that it pushed the ear out of position – nothing was noticeable. Also Lulu had displayed no signs of distress, which was amazing when one considered the size of the lump. She allowed me to explore the exact extent of the swelling gently, without doing anything more than carefully and politely removing my fingers if their pressure became too painful. I decided, after investigation, that I would have to lance it, as it was obviously full of matter, so I picked Lulu up in my arms and carried her into the house, where I put her down on the sofa and gave her a banana to keep her occupied until I had everything ready.

Up till now, the chimps had only been allowed in the house on very special occasions, and Lulu was, therefore, charmed with the idea that she was getting an extra treat without Chumley's knowledge. She sat on the sofa, her mouth full of banana, giving a regal handshake and a muffled hoot of greeting to whoever came into the room, rather as though she owned the place and you were attending one of her 'At Homes'. Presently, when everything was ready, I sat down beside her on the sofa and gently cut away the long hairs behind the ear that was affected. When it was fully exposed, the swelling looked even worse than before, a rich plum colour, and the skin had a leathery appearance. I carefully swabbed the whole area with disinfected warm water, searching to see if I could find a head or an opening to the swelling, for I was now convinced that it was a boil or ulcer that had become infected, but I could find no opening at all. Meanwhile, Lulu, having carefully and thoroughly scrutinized all the medical paraphernalia, had devoted her time to consuming another banana. I took a hypodermic needle and gently pricked the discoloured skin all over the swelling without causing her to deviate from the paths of gluttony, so it was obvious that the whole of the discoloured area was dead skin.

I was faced with something of a problem. Although I was fairly sure that I could make an incision across the dead skin, and thus let out the pus, without Lulu suffering any pain, I was not absolutely certain about it. She was, as I have remarked, of a lovable and charming disposition, but she was also a large, well-built ape, with a fine set of teeth, and I had no desire to enter into a trial of strength with her. The thing to do was to keep her mind occupied elsewhere while I tackled the job, for, like most chimps, Lulu was incapable of thinking of more than one thing at a time. I enlisted the aid of my mother and

Jacquie, to whom I handed a large tin of chocolate biscuits, with instructions that they were to feed them to Lulu at intervals throughout the ensuing operation. I had no fears for their safety, as I knew that if Lulu was provoked into biting anyone it would be me. Uttering up a brief prayer, I sterilized a scalpel, prepared cotton wool swabs, disinfected my hands and went to work. I drew the scalpel blade across the swelling, but, to my dismay, I found that the skin was as tough as shoe-leather, and the blade merely skidded off. I tried a second time, using greater pressure, but with the same result. Mother and Jacquie kept up a nervous barrage of chocolate biscuits, each of which was greeted with delighted and slightly sticky grunts from Lulu.

'Can't you hurry up?' inquired Jacquie, 'these biscuits won't last for ever.'

'I'm doing the best I can,' I said irascibly, 'and a nurse doesn't tell a doctor to hurry up in the middle of an operation.'

'I think I've got some chocolates in my room, dear,' said my mother helpfully, 'shall I fetch them?'

'Yes, I should, just in case.'

While Mother went off to fetch the chocolates, I decided that the only way to break into the swelling was to jab the point of the scalpel in and then drag it downwards, and this I did. It was successful: a stream of thick putrid matter gushed out from the incision, covering both me and the sofa. The smell from it was ghastly, and Jacquie and Mother retreated across the room hastily. Lulu sat there, quite unperturbed, eating chocolate biscuits. Endeavouring not to breathe more often than was necessary, I put pressure on the swelling, and eventually, when it was empty, I must have relieved it of about half a cupful of putrefying blood and pus. With a pair of scissors I carefully clipped away the dead skin and disinfected the raw area that was left. It was useless trying to put a dressing on, for I knew that Lulu would remove it as soon as she was put back in her cage. When I had cleaned it up to my satisfaction, I picked Lulu up in my arms and carried her back to her cage. Here she greeted Chumley with true wifely devotion, but Chumley was deeply suspicious. He examined her ear carefully, but decided that it was of no interest. Then, during one of Lulu's hoots of pleasure, leant forward and smelt her breath. Obviously, she had been eating chocolate, so Lulu, instead of receiving a husbandly embrace, received a swift clout over the back of the head. In the end, I had to go and fetch the rest of the chocolate biscuits to placate Chumley. Lulu's ear healed up perfectly, and within six months you had to look very closely to see the scar.

About a year later Chumley decided that it was his turn to fall ill, and of course he did it – as he did all things – in the grand manner. Chumley, I was told, had tooth ache. This rather surprised me, as, not long before, he had lost his baby teeth and acquired his adult ones, and I thought it was a bit too soon for any of them to have decayed. Still, there he was, squatting forlornly in the cage, clasping his jaw and ear with his hand and looking thoroughly miserable. He was obviously in pain, but I was not sure whether it was his ear or his jaw that was the cause of it. The pain must have been considerable, for he would not let me take his hand away to examine the side of his face, and when I persisted in trying he became so upset that it was clear I was doing more harm than good, so I had to give up. I stood for a long time by the cage, trying to deduce from his actions what was the matter with him. He kept lying down, with the bad side of his head cuddled by his hand, and whimpering gently to himself; once, when he had climbed up the wire to relieve himself, he lowered himself to the ground again rather awkwardly, and as his feet thumped on to the floor of the cage he screamed, as though the jar had caused him considerable pain. He refused all food and, what

was worse, he refused all liquids as well, so I could not give
him any antibiotics. We had to remove Lulu, as, instead of
showing wifely concern, she bounded round the cage,

occasionally bumping into Chumley, or leaping on to him and making him cry out with pain.

I became so worried about his condition by the afternoon that I called into consultation Mr Blampier, a local veterinary surgeon, and our local doctor. The latter, I think, was somewhat surprised that he should be asked to take a chimpanzee on to his panel, but agreed nevertheless. It was plain that Chumley's jaw and ear would have to be examined carefully, and I knew that, in his present state, he would not allow that, so it was agreed that we would have to anaesthetize him. This is what had to be done, but how to do it was another matter. Eventually, it was decided that I should try to give Chumley an injection of a tranquillizer which would, we hoped, have him in an agreeable frame of mind by the evening to accept an anaesthetic. The problem was whether or not Chumley was going to let me give him the injection. He was lying huddled up in his bed of straw, his back towards me, and I could see he was in great pain, for he never even looked round to see who had opened the door of his cage. I talked to him, in my best bedside manner, for a quarter of an hour or so, and at the end of that time he was allowing me to stroke his back and legs. This was a great advance, for up till now he had not let me come within stroking distance. Then, plucking up my courage, and still talking feverishly, I picked up the hypodermic and swiftly slipped the needle into the flesh of his thigh. To my relief, he gave no sign of having noticed it. As gently and as slowly as I could, I pressed the plunger and injected the tranquillizer. He must have felt this, for he gave a tiny, rather plaintive hoot, but he was too apathetic to worry about it. Still, talking cheerful nonsense, I closed the door of his bedroom and left the drug to take effect.

That evening Dr Taylor and Mr Blampier arrived, and I reported that the tranquillizer had taken effect: Chumley was in a semi-doped condition, but, even so, he still would not let me examine his ear. So we repaired to his boudoir, outside of which I had rigged up some strong lights and a trestle table on which to lay our patient. Doctor Taylor poured ether on to a mask, and I opened Chumley's bedroom door, leant in and placed the mask gently over his face. He made one or two half-hearted attempts to push it away with his hand, but the ether, combined with the tranquillizer was too much for him, and he slipped into unconsciousness quite rapidly. As soon as he was completely under, we hauled him out of the cage and laid him on the trestle table, still keeping the mask over his face. Then the experts went to work. First, his ear was examined, and found to be perfectly healthy; just for good measure, we examined his other ear as well, and that, too, was all right. We then opened his mouth and carefully checked his teeth: they were an array of perfect, glistening white dentures without a speck of decay on any of them. We examined his cheeks, his jaw and the whole of his head, and could not find a single thing wrong. We looked at his neck and shoulders, with the same result. As far as we could ascertain, there was nothing the matter with Chumley whatsoever, and yet *something* had been causing him considerable pain. Dr Taylor and Mr Blampier departed, much mystified, and I carried Chumley into the house, wrapped in a blanket, and put him on a campbed in front of the drawing-room fire. Then Jacquie brought more blankets, which we piled on top of him, and we sat down to wait for the anaesthetic to wear off.

Lying there, his eyes closed, breathing out ether fumes stertorously, he looked like a slightly satanic cherub who, tired out after a day's mischief-making, was taking a well-earned rest. The amount of ether he was expelling from his lungs made the whole room reek, so that we were forced to open a window. It was about half an hour before he began to sigh deeply and twitch, as a preliminary to regaining consciousness, and I went over and sat by the bedside with a cup of water ready, since I knew from experience the dreadful thirst that assails one when you come out from under an anaesthetic. In a few minutes he opened his eyes, and as soon as he saw me he gave a feeble hoot of greeting and held out his hand, in spite of the fact that he was still half asleep. I held up his head and put the cup to his lips and he sucked at the water greedily before the ether overcame him again and he sunk back into sleep. I decided that an ordinary cup was too unwieldy to give him drinks, as a considerable quantity of liquid was spilt. I managed, by ringing up my friends, to procure an invalid's cup, one of those articles that resemble a deformed teapot, and the next time Chumley woke up this proved a great success, as he could suck water out of the spout without having to sit up.

Although he recognized us, he was still in a very drugged and stupid state, and so I decided that I would spend the night sleeping on the sofa near him, in case he awakened and wanted anything. Having given him another drink, I made up my bed on the sofa, turned out the light and dozed off. About two o'clock in the morning I was awakened by a crash in the far corner of the room. I hastily put on the light to find that Chumley was awake and wandering round the room, like a drunken man, barging into all the furniture. As soon as the light came on and he saw me, he uttered a scream of joy, staggered across the room and insisted on embracing and kissing me before gulping down a vast drink of water. I then helped him back on to his bed and covered him with his blankets, and he slept peacefully until daylight.

He spent the day lying quietly on his bed, staring up at the ceiling. He ate a few grapes and drank great quantities of glucose and water, which was encouraging. The most encouraging thing, however, was that he no longer held the side of his face and did not appear to be suffering any pain. In some extraordinary way we seemed to have cured him without doing anything. When Dr Taylor telephoned later that day to find out how Chumley was faring I explained this to him, and he was as puzzled as I. Then, later on, he rang up to say that he had thought of a possible explanation: Chumley might have been suffering from a slipped disc. This could have caused intense pain in the nerves of jaw and ear, without there being anything externally to show what caused it. When we had Chumley limp and relaxed under the anaesthetic we pulled his head around quite a lot during our examination, and probably pushed the disc back into place, without realizing it. Mr Blampier agreed with this diagnosis. We had no proof, of course, but certainly Chumley was completely cured, and there was no recurrence of the pain. He had naturally lost a lot of weight during his illness, and so for two or three weeks he was kept in a specially heated cage and fed up on every delicacy. Within a very short time he had put on weight and was his old self, so that whenever anyone went near his cage they were showered with handfuls of sawdust. This, I presume, was Chumley's way of thanking one.

Sometimes animals injure themselves in the most ridiculous way imaginable. Hawks and pheasants, for example, are the most hysterical of birds. If anything unusual happens they get into a terrible state, fly straight up, like rockets, and crash into the roof of their cage, either breaking their necks or neatly scalping themselves. But there are other birds equally stupid. Take the case of Samuel.

Samuel is a South American seriama. They are not unlike the African secretary bird. About the size of a half-grown turkey, they have long, strong legs, and a ridiculous little tuft of feathers perched on top of their beaks. In the wild state seriamas do not fly a great deal, spending most of their time striding about the grasslands in search of snakes, mice, frogs and other delicacies. I had purchased Samuel from an Indian in Northern Argentina, and as he had been hand-reared he was,

of course, perfectly (and sometimes embarrassingly) tame. When I finally shipped him back to Jersey with the rest of the animals, we took him out of his small travelling crate and released him in a nice, spacious aviary. Samuel was delighted, and to show us his gratitude the first thing he did was to fly up on to the perch, fall off it and break his left leg. There are times when animals do such idiotic things that you are left bereft of words.

Fortunately, for Samuel, it was a nice clean break, half-way down what would be the shin in a human being. We made a good job of splinting it, covering the splint with plaster of Paris bandage, and, when it was dry, put him in a small cage so that he could not move around too much. The following day his foot was slightly swollen, so I gave him a penicillin injection – which he took great exception to – and his foot returned to normal size as a result. When we eventually took off the splint, we found the bones had knitted perfectly, and to-day, as he strides importantly around his aviary, you have to look very closely to see which leg it was that he broke. Knowing Samuel for the imbecile he is, it would not surprise me in the least if he did not repeat the performance at some time in the future . . . probably on a day when I am up to my eyes in other work.

During the course of your Florence Nightingale work you become quite used to being bitten, scratched, kicked and bruised by your patients, and on many occasions, having performed first-aid on them, you have to perform it on yourself. Nor is it always the bigger creatures that are the most dangerous to deal with. A squirrel or a pouched rat can inflict almost as much damage as a flock of Bengal tigers when they put their minds to it. While anointing a fluffy, gooey-eyed bush-baby once for a slight skin infection on the tail, I was bitten so severely in the thumb that it went septic, and I had to have it bandaged for ten days. The bushbaby was cured in forty-eight hours.

Human doctors are covered by the Hippocratic Oath. The wild animal doctor employs a variety of oaths, all rich and colourful, but which would, I feel, be frowned upon by the British Medical Council.

CHAPTER SIX

LOVE AND MARRIAGE

Dear Mr Durrell,
I am seven years old and I have just had a baby tortoise . . .

You can tell if an animal is happy in captivity in a number of different ways. Principally, you can tell by its condition and appetite, for a creature which has glossy fur or feathering, and eats well to boot, is obviously not pining. The final test that proves beyond a shadow of doubt that the animal has accepted its cage as 'home' is when it breeds.

At one time, if an animal did not live very long in captivity, or did not breed, the zoos seemed to be under the impression that there was something wrong with it, and not something wrong with their methods of keeping it. So-and-so was '*impossible*' to keep in captivity, they would say, and, even if it did manage to survive for a while, it was '*impossible*' to breed. These sweeping statements were delivered in a wounded tone of voice, as if the wretched creature had entered into some awful conspiracy against you, refusing to live or mate. At one time there was a huge list of animals that, it was said, were impossible to keep or breed in confinement, and this list included such things as the great apes, elephants, rhinoceros, hippopotamus and so on. Gradually, over the years, one or two more agile brains entered the zoo world, and to everyone's surprise and chagrin it was discovered that the deaths and lack of babies were not due to stubbornness on the part of the creature, but due to lack of knowledge and experiment on the part of the people who kept them. I am convinced there are precious few species of animals which you cannot successfully maintain *and* breed, once you have found the knack. And by knack I mean once you have discovered the right type of caging, the best-liked food, and, above all, a suitable mate. On the face of it, this seems simple enough, but it may take several years of experiment before you acquire them all.

Marriages in zoos are, of course, arranged, as they used to be by the eighteenth-century Mammas. But the eighteenth century Mamma had one advantage over the zoo: having married off her daughters, there was an end to it. In a zoo you are never quite sure, since any number of things may happen. Before you can even lead your creatures to the altar, so to speak, it is quite possible that either the male or the female may take an instant dislike to the mate selected, and so, if you are not careful, the bride or groom may turn into a corpse long before the honeymoon has started. A zoo matchmaker has a great number of matters to consider, and a great number of risks to take, before he can sit back with a sigh of relief and feel the marriage is an accomplished fact. Let us take the marriage of Charles, as a fairly typical one.

Charles is – rather unzoologically – what is known as a Rock ape from Gibraltar. He is, of course, not an ape at all but a Macaque, one of a large group of monkeys found in the Far East. Their presence in North Africa is puzzling, but obviously they have been imported to the Rock of Gibraltar and have thus gained the doubtful distinction of being the only European monkey. We were offered Charles when the troop on the Rock underwent its periodical thinning, and we were very pleased to have him. He was brought over from Gibraltar in style on one of Her Majesty's ships, and we duly took possession of him. He stood two feet six inches high when squatting on his haunches, and was clad in an immensely long, thick, gingery brown coat. His walk was very dog-like but with a distinct swagger to it, as befits a member of the famous Rock garrison. He had bright, intelligent brown eyes and a curious pale pinkish face, thickly covered with freckles

He was undoubtedly ugly, but with an ugliness that was peculiarly appealing. Curiously enough, although he was a powerful monkey, he was excessively timid, and an attempt to keep him with a mixed group of other primates failed, for they bullied him unmercifully. So Charles was moved to a cage of his own, and a carefully worded letter was dispatched to the Governor of Gibraltar, explaining in heart-rending terms, Charles' solitary confinement, and hinting that he would be more than delighted if a female Rock ape should be forthcoming. In due course we received a signal to say that Charles' condition of celibacy had been reviewed and it had been decided that, as a special concession, a female Rock ape, named Sue, was going to be sent to us. Thus another of Her Majesty's ships was pressed into service, and Sue duly arrived.

By this time, of course, Charles had settled down well in his new cage, and had come to look upon it as his own territory, and so we had no idea how he would treat the introduction of a new Rock ape – even a female one – into his bachelor apartments. We carried Sue in her travelling crate and put it on the ground outside Charles' cage, so that they could see each other. Sue became very excited when she saw him, and chattered away loudly, whereas Charles, after the first astonished glance, sat down and stared at her with an expression of such loathing and contempt on his freckled face that our hearts sank. However, we had to take the plunge, and Sue was let into the cage. She sprang out of her crate with great alacrity, and set off to explore the cage. Charles, who had been sitting up in the branches dissociating himself from the whole procedure up till then, decided the time had come to assert himself. He leapt down to the ground and sprang on Sue before she realized what was happening and could take evasive action. Within a second she had received a sharp nip on the shoulder, had her hair pulled and her ears boxed, and was sent tumbling into a

corner of the cage. Charles was back on his branch, lookin
around with a self-satisfied air, uttering little grunts to hin
self. We hastily went and fetched two big bowls of fruit an
put them into the cage, whereupon Charles came down an
started to pick them over with the air of a gourmet, while Su
sat, watching him hungrily. Eventually, the sight of th
grape juice trickling down Charles' chin was too much for he
and she crept forward timidly, leant towards the bowl an
took a grape, which she hastily crammed into her mouth, i
case Charles went for her. He completely ignored her, how
ever, after one quick glance from under his eyebrows, an
gaining courage, she again leant forward and grabbed a whol
handful of grapes. Within a few minutes they were both fee
ing happily out of the same dish, and we sighed with relie
An hour later, when I passed by, there was Charles, lying o
his back, eyes closed, a blissful expression on his face, whil
Sue, with a look of deep concentration, was searching his fu
thoroughly. It seemed that his original attack on Sue wa
merely to tell her that it was *his* cage, and that, if she wante
to live there, she had to respect his authority.

Sometimes one acquires mates for animals in very curiou
ways. One of the most peculiar was the way in which we foun
a husband for Flower. Now, Flower was a very handsom
North American skunk, and when she first came to us she wa
slim and sylph-like and very tame. Unfortunately, Flowe
decided that there were only two things in life worth doing
eating and sleeping. The result of this exhausting life, whic
she led, was that she became so grossly overweight that sh
was – quite literally – circular. We tried dieting her, but wit
no effect. We became somewhat alarmed, for overweight ca
kill an animal as easily as starvation. It was plain that wha
Flower needed was exercise, and equally plain that she had n
intention of going out of her way to obtain it. We decide
that what she needed was a mate, but, at that particular tim

skunks were in short supply and none were obtainable, so Flower continued to eat and sleep undisturbed. Then, one day, Jacquie and I happened to be in London on business, and, being a bit early for our appointment, we walked to our destination. On rounding a corner, we saw approaching us a little man dressed in a green uniform with brass buttons, carrying in his arms – above all things – a baby chimpanzee. At first, with the incongruous combination of the uniform and the ape, we were rather taken aback, but as he came up to us I recovered my wits and stopped him.

'What on earth are you doing with a chimpanzee?' I asked him, though why he should not have a perfect right to walk through the streets with a chimpanzee I was not quite sure.

'I works for Viscount Churchill,' he explained, 'and he

keeps a lot of queer pets. We've got a skunk, too, but we'll 'ave to get rid of that, 'cos the chimp don't like it.'

'A skunk?' I said eagerly. 'Are you sure it's a skunk?'

'Yes,' replied the little man, 'positive.'

'Well, you've met just the right person,' I said. 'Will you give my card to Viscount Churchill and tell him that I would be delighted to have his skunk, if he wants to part with it?'

'Sure,' replied the little man, 'I should think he'd be pleased to let you 'ave it.'

We returned to Jersey full of hope that we might have found a companion, if not a mate, for Flower. Within a few days I received a courteous letter from Viscount Churchill, saying that he would be very pleased to let his skunk come to us, and that, as soon as he had had a travelling cage constructed, he would send him. The next thing I received was a telegram. Its contents were simple and to the point, but I cannot help feeling that it must have puzzled the postal authorities. It read as follows:

GERALD DURRELL ZOOLOGICAL PARK LES AUGRES JERSEY CI: GLADSTONE LEAVING FLIGHT BE 112 AT 19 HOURS TODAY THURSDAY CAGE YOUR PROPERTY.   CHURCHILL.

Gladstone, on being unpacked, proved to be a lovely young male, and it was with great excitement that we put him in with Flower and stood back to see what would happen. Flower was, as usual, lying in her bed of straw, looking like a black and white, fur-covered football. Gladstone peered at this apparition somewhat short-sightedly and then ambled over to have a closer look. At that moment Flower had one of her brief moments of consciousness. During the day she used to wake up periodically for about thirty seconds at a time, just long enough to have a quick glance round the cage to see if anyone had put a plate of food in while she slept. Gladstone, suddenly perceiving that the football had a head, stopped in astonishment and put up all his fur defensively. I am quite sure that for a moment he was not certain what Flower *was*, and I can hardly say I blame him, for when she was just awakened from a deep sleep like that she rarely looked her best. Gladstone stood staring at her, his tail erect like an exclamation mark; Flower peered at him blearily and, because he was standing so still and because she had a one track mind, Flower obviously thought he was some new and exotic dish which had been put in for her edification. She hauled herself out of her bed and waddled across towards Gladstone.

Flower walking looked, if anything, more extraordinary than Flower reclining. You could not see her feet, and so you had the impression of a large ball of black and white fur propelling itself in your direction in some mysterious fashion. Gladstone took one look, and then his nerve broke and he ran and hid in the corner. Flower, having discovered that he was only a skunk, and therefore not something edible, retreated once more to her bed to catch up on her interrupted nap. Gladstone steered clear of her for the rest of the day, but towards evening he did pluck up sufficient courage to go and sniff her sleeping form and find out *what* she was, a discovery that seemed to interest him as little as it had done Flower. Gradually, over a period of days, they grew very fond of one another, and then came the great night when I passed their cage in bright moonlight and was struck dumb with astonishment, for there was Gladstone chasing Flower round and round the cage, and Flower (panting and gasping for breath) was actually enjoying it. When he at length caught her, they rolled over and over in mock battle, and, when they had finished, Flower was so out of breath she had to retire to bed for a short rest. But this was only the beginning, for after a few months of Gladstone's company Flower regained her girlish figure, and before long she could out-run and out-wrestle Gladstone himself.

So zoo marriages can be successful or unsuccessful, but if they are successful they should generally result in some progeny, and this again presents you with further problems. The most important thing to do, if you can, is to spot that a happy event is likely to take place as far in advance as possible, so that the mother-to-be can be given extra food, vitamins and so forth. The second most important thing is to make up your mind about the father-to-be: does he stay with the mother, or not? Fathers, in fact, are sometimes more of a problem than the mothers. If you do not remove them from the cage, they might worry the female, so that she may give birth prematurely; on the other hand, if you do remove them, the female may pine and again give birth prematurely. If the father is left in the cage, he might well become jealous of the babies and eat them; on the other hand, he might give the female great assistance in looking after the young: cleaning them and keeping them amused. So, when you know that a female is pregnant, one of your major problems is what to do with Dad, and at times, if you do not act swiftly, a tragedy may occur.

We had a pair of slender loris of which we were inordinately proud. These creatures look rather like drug-addicts that have seen better days. Clad in light grey fur they have enormously long and thin limbs and body; strange, almost human, hands; and large lustrous brown eyes, each surrounded by a circle of dark fur, so that the animal appears as though it is either recovering from some ghastly debauchery, or a very unsuccessful boxing tournament. They have a reputation for being extremely difficult and delicate to keep in captivity, which, by and large, seems to be true. This is why we were so proud of our pair, as we had kept them for four years, and this was a record. By careful experiment and observation, we had worked out a diet which seemed to suit them perfectly. It was a diet that would not have satisfied any other creature *but* a slender loris, consisting as it did of banana, mealworms and milk, but nevertheless on this monotonous fare they lived and thrived.

As I say, we were very proud that our pair did so well, so you can imagine our excitement when we realized that the female was pregnant: this was indeed going to be an event, the first time slender loris had been bred in captivity, to the best of my knowledge. But now we were faced with the father problem, as always; and, as always, we teetered. Should we remove him or not? At last, after much deliberation, we decided not to do so, for they were a very devoted couple. The great day came, and a fine, healthy youngster was born. We put up screens round the cage, so that the parents would not be disturbed by visitors to the Zoo, gave them extra titbits, and watched anxiously to make sure the father behaved himself. All went well for three days, during which time the parents kept close together as usual, and the baby clung to its mother's fur with the tenacity and determination of a drowning man clasping a straw. Then, on the fourth morning, all our hopes were shattered. The baby was lying dead at the bottom of the cage, and the mother had been blinded in one eye by a savage bite on the side of her face. To this day we do not know what happened, but I can only presume that the male wanted to mate with the female, and she, with the baby clinging to her, was not willing, and so the father turned on her. It was a bitter blow, but it taught us one thing: should we ever succeed in breeding slender loris again, the father will be removed from the cage as soon as the baby is born.

In the case of some animals, of course, removing the father would be the worst thing you could do. Take the marmosets, for instance. Here the male takes the babies over the moment they are born, cleans them, has them both clinging to his body and only hands them over to the mother at feeding time. I had wanted to observe this strange process for a long time, and

thus, when one of our cotton-eared marmosets became pregnant, I was very pleased. My only fear was that she would give birth to the baby when I happened to be away, but luckily this did not happen. Instead, very early one morning, Jeremy burst into my bedroom with the news that he thought the marmoset was about to give birth, so, hastily flinging on some clothes, I rushed down to the Mammal House. There I found the parents to be both unperturbed, clinging to the wire of their cage and chittering hopefully at any human who passed. It was quite obvious from the female's condition that she would give birth fairly soon, but she seemed infinitely less worried by the imminence of this event than I. Getting myself a chair, I sat down to watch. I stared at the female marmoset, and she stared at me, while in the corner of the cage her husband – with typical male callousness – sat stuffing himself on grapes and mealworms, and took not the slightest notice of his wife.

Three hours later there was absolutely no change, except that the male marmoset had finished all the grapes and mealworms. By then my secretary had arrived and, as I had a lot of letters to answer, I made her bring a chair and sit down beside

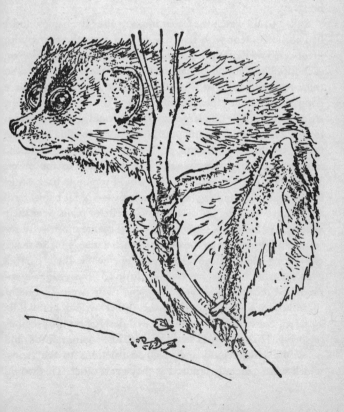

me in front of the marmoset cage while I dictated. I think that visitors to the Zoo that day must have thought slightly eccentric the sight of a man dictating letters, while keeping his eyes fixed hypnotically on a cageful of marmosets. Then, about midday, someone arrived whom I had to see. I was away from the cage, for approximately ten minutes, and on my return the father marmoset was busy washing two tiny scraps of fur that were clinging to him vigorously. I could quite cheerfully have strangled the female marmoset: after all my patient waiting, she went and gave birth during the short period I happened to be away.

Still, I could watch the father looking after the babies, and I had to be content with that. He looked after the twins with great care and devotion, generally carrying them slung one on each hip, like a couple of panniers on a donkey. His fur was so thick and the babies so small that most of the time they were completely hidden; then, suddenly, from the depths of his fur, a tiny face, the size of a large hazel nut, would appear, and two bright eyes would regard you gravely. At feeding time the father would go and hang on the wire close alongside the mother, and the babies would pass from one to the other. Then, their thirst quenched, they would scramble back on to father again. The father was extremely proud of his babies, and was always working himself up into a state of panic over their welfare. As the twins grew older, they became more venturesome, and would leave the safety of their father's fur to make excursions along the nearby branches, while their parents eyed them with pride, as well as a little anxiously. If you approached too near the cage when the twins were on one of their voyages of exploration, the father would get wildly agitated, convinced that you had evil designs on his precious offspring. His fur would stand on end, like an angry cat's, and he would chitter loud and shrill instructions to the twins, which were generally ignored as they grew older. This would reduce him to an even worse state of mind, and screaming with rage and fear, he would dive through the branches, grab the twins and sling them into place, one on each hip; then, muttering dark things to himself – presumably about the disobedience of the modern generation – he would potter off to have a light snack to restore his nerves, casting dark glances at you over his shoulder. Watching the marmoset family was an enchanting experience, more like watching a troupe of strange little fur-covered leprechauns than monkeys.

Naturally, the biggest thrill is when you succeed in breeding some creature which you know from the start is going to be extremely difficult. During my visit to West Africa I had managed to acquire some Fernand's skinks, probably one of the most beautiful of the lizard family, for their big, heavy bodies are covered with a mosaic work of highly polished scales in lemon yellow, black, white and vivid cherry-red. By the time the Zoo in Jersey was established I had only two of these magnificent creatures left, but they were fine, healthy specimens, and they settled down well in the Reptile House. Sexing most reptiles is well-nigh impossible, so I did not know if these skinks were a true pair or not, but I did know that, even if by some remote chance they were a true pair, the chances of breeding them were one in a million. The reason for this lay in the fact that reptiles, by and large, lay the most difficult eggs to hatch in captivity. Tortoises, for example, lay hard-shelled eggs which they bury in earth or sand. But if you do not get the temperature and humidity just right in the cage, the eggs will either become mildewed or else the yolk will dry up. A lot of lizards, on the other hand, lay eggs which have a soft, parchment-like shell, which makes matters a bit more difficult, for they are even more sensitive to moisture and temperature.

Knowing all this, I viewed with mixed feelings the clutch of a dozen eggs which the female Fernand's skink laid one morning in the earth at the bottom of her cage. They were white, oval eggs, each about the size of a sugared almond, and the female (as happens among some of the skinks) stood guard over them and would attack your hand quite fearlessly should you put it near the eggs. With most lizards the female walks off, having laid her batch, and forgets all about it; in the case of some of the skinks, however, the female guards the nest and, lying on top of the soil in which the eggs are buried, urinates over the nest at intervals to maintain the right moisture content, in order to keep the delicate shells from shrivelling up in the heat. Our female skink appeared to know what she was doing, and so all we could do was to sit back and await developments, without any great hope that the eggs would hatch. As week after week passed, our hopes sank lower and lower, until, eventually, I dug down to the nest, expecting to find every egg shrivelled up. To my surprise, however, I found that only four eggs had done so; the rest were still plump and soft, though discoloured, of course. I removed the four shrivelled ones and carefully opened one with a scalpel. I found a dead but well-developed embryo. This was encouraging, for it proved at least that the eggs were fertile. So we sat back to wait again.

Then, one morning, I was down in the Reptile House about some matter, and as I passed the skink's cage, I happened to glance inside. As usual, the cage looked empty, as the parent skinks spent a lot of their time buried in the soil at the bottom of the cage. I was just about to turn away when a movement among the dry leaves and moss attracted my attention. I peered more closely to see what had caused it, and suddenly, from around the edge of a large leaf, I saw a minute pink and black head peering at me. I could hardly believe my eyes, and stood stock still and stared as this tiny replica of the parents slowly crept out from behind the leaf. It was about an inch and a half long, with all the rich colouring of the adult, but so slender, fragile and glossy that it resembled one of those ornamental brooches that women wear on the lapels of their coats. I decided that if one had hatched, there might be others, and I wanted to remove them as quickly as possible for, although the female had been an exemplary mother till now, it was quite possible that either she or the male might eat the youngsters.

We prepared a small aquarium and very carefully caught the baby skink and put him into it. Then we set to work and stripped the skink's cage. This was a prolonged job, for each leaf, each piece of wood, each tuft of grass had to be checked and double checked, to make sure there was no baby skink curled up in it. When the last leaf had been examined, we had four baby Fernand's skinks running around in the aquarium. When you consider the chances of any of the eggs hatching at all, to have four out of twelve was, I thought, no mean feat. The only thing that marred our delight at this event was that the baby skinks had decided to hatch at the beginning of the winter, and as they could only feed off minute things the job of finding them enough food was going to be difficult. Tiny mealworms were, of course, our standby, but all our friends with gardens rallied round, and would come up to the Zoo once or twice a week, bearing biscuit-tins full of woodlice, earwigs, tiny snails and other dainty morsels that gave the babies the so necessary variety in diet. Thus the tiny reptiles thrived and grew. At the time of writing, they are about six inches long, and as handsome as their parents. I hope it will not be long before they start laying eggs, so that we can try to rear a second generation in captivity.

There are, of course, some animals which could only with the greatest difficulty be prevented from breeding in captivity, and among these are the coatimundis. These little South American animals are about the size of a small dog, with long, ringed tails which they generally carry pointing straight up in the air. They have short, rather bowed legs, which give them a bear-like, rolling gait; and long, rubbery, tip-tilted noses which are forever whiffling to and fro, investigating every nook and cranny in search of food. They come in two colours: a brindled greenish-brown, and a rich chestnut. Martha and Mathias, the pair I had brought back from Argentina, were of the brindled kind.

As soon as these two had settled down in their new cage in the Zoo they started to breed with great enthusiasm. We noticed some very interesting facts about this which are worth recording. Normally, Mathias was the dominant one. It was he who went round the cage periodically 'marking' with his scent gland so that everyone would know it was his territory. He led Martha rather a dog's life, pinching all the best bits of food until we were forced to feed them separately. This Victorian male attitude was only apparent when Martha was not pregnant. As soon as she had conceived, the tables were turned. She was now the dominant one, and made poor Mathias's life hell, attacking him without provocation, driving him away from the food, and generally behaving in a very shrewish fashion. It was only by watching to see which was the dominant one at the moment, that we could tell, in the early stages, whether Martha was expecting a litter or not.

Martha's first litter consisted of four babies, and she was very proud of them, and proved to be a very good mother indeed. We were not sure what Mathias's reactions to the youngsters was going to be, so we had constructed a special shut-off for him, from which he could see and smell the babies without being able to sink his teeth into them, should he be so inclined. It turned out later that Mathias was just as full of pride in them as Martha, but in the early stages we were not taking any risks. Then the great day came when Martha considered the babies were old enough to be shown to the world,

and so she led them out of her den and into the outside cage
for a few hours a day. Baby coatimundis are, in many ways, the
most enchanting of young animals. They appear to be all
head and nose, high-domed, intellectual-looking foreheads,
and noses that are, if anything, twice as rubbery and inquisitive as the adults'. Then they are natural clowns, forever
tumbling about, or sitting on their bottoms in the most
human fashion, their hands on their knees. All this, combined
with their ridiculous rolling, flat-footed walk, made them quite
irresistible. They would play follow-my-leader up the
branches in their cage, and when the leader had reached the
highest point he would suddenly go into reverse, barging into
the one behind, who, in his turn, was forced to back into the
one behind him, and so on, until they were all descending the
branch backwards, trilling and twittering to each other
musically. Then they would climb up into the branches and
do daring trapeze acts, hanging by their hind paws, or one
fore paw, swinging to and fro, trying their best to knock each
other off. Although they often fell from quite considerable
heights on to the cement floor, they seemed as resilient as
india-rubber and never hurt themselves.

When they grew a little older and discovered they could
squeeze through the wire mesh of the cage, they would escape
and play about just inside the barrier rail. Martha would keep
an anxious eye on them during these excursions, and should
any real or imaginary danger threaten they would come
scampering back at her alarm cry, and, panting excitedly,
squeeze their fat bodies through the wire mesh and into the
safety of the cage. As they grew bolder, they took to playing
farther and farther afield. If there were only a few visitors
about, they would go and have wrestling matches on the main
drive that sloped down past their cage. In many ways this
was a nuisance, for at least twenty times a day some kindly
and well-intentioned visitor would come panting up to us

with the news that some of our animals were out, and we would have to explain the whole coatimundi set-up.

It was while the babies were playing on the back drive one day that they received a fright which had a salutary effect on them. They had gradually been going farther and farther away from the safety of their cage, and their mother had been growing increasingly anxious. The babies had just learnt how to somersault, and were in no mood to listen to their mother's warnings. It was when they had reached a point quite far from their cage that Jeremy drove down the back drive in the Zoo van. Martha uttered her warning cry, and the babies, stopping their game, suddenly saw they were about to be attacked by an enormous roaring monster that was between them and the safety of their home. Panic-stricken they turned and ran. They galloped flat-footedly down past the baboon cage, past the chimps' cage, past the bear cage, without finding anywhere to hide from the monster that pursued them. Suddenly, they saw a haven of safety, and the four of them dived for it. The fact that the ladies' lavatory happened to be empty at that moment was entirely fortuitous. Jeremy, cursing all coatimundis, crammed on his brakes and got out. He glanced round surreptitiously to make sure there were no female visitors around, and then dived into the ladies' in pursuit of the babies. Inside, they were nowhere to be seen, and he was just beginning to wonder where on earth they had got to when muffled squeaks from inside one of the cubicles attracted him. He discovered that all four babies had squeezed under the door of one of these compartments. What annoyed Jeremy most of all, though, was that he had to put a penny in the door *to get them out*.

Still, whatever tribulations they might give you, the babies in the Zoo provide tremendous pleasure and satisfaction. The sight of the peccaries playing wild games of catch-as-catch-can with their tiny piglets; the baby coatimundis rolling and

bouncing like a circus troupe; the baby skinks in their miniature world, carefully stalking an earwig almost as big as themselves; the baby marmosets dancing through the branches, like little gnomes, hotly pursued by their harassed father: all these things are awfully exciting. After all, there is no point in having a Zoo unless you breed the animals in it, for by breeding them you know that they have come to trust you, and that they are content.

CHAPTER SEVEN

A GORILLA IN THE GUEST-ROOM

Dear Mr Durrell,
Could you please have our Rhesus monkey? He is growing so big and
jumping on us from trees and doing damage and causing so much
trouble. Already my mother has been in bed with the doctor three
times ...

IT was towards the end of the second year that I decided, the
Zoo now being well established, it must cease just being a
mere showplace of animals, and start to contribute something
towards the conservation of wildlife. I felt that it would be
essential to gradually weed out all the commoner animals in
the collection and to replace them with rare and threatened
species. That is, species which were threatened with extinction
in the wild state. The list of these was a long and melancholy
one; in fact – without reptiles – it filled three fat volumes. I
was wondering which of this massive list of endangered
species we could start with, when the decision was taken out
of my hands. An animal dealer telephoned and asked me if I
wanted a baby gorilla.

Gorillas have never been exceptionally numerous as a
species and with the state that Africa was in (politically speak-
ing) at that moment it seemed to me that they might well
become extinct within the next twenty years, for newly
emergent governments are generally far too busy proving
themselves to the world for the first few years to worry much
about the fate of the wild-life of their country, and history
has proved, time and time again, how rapidly a species can be
exterminated, even a numerous one. So the gorilla had been
high on my list of priorities. I was not convinced, however,

that the animal in question really was a gorilla. In my experience, the average animal dealer can, with difficulty, distinguish between a bird, a reptile and a mammal, but this is about the extent of his zoological knowledge. I felt that it was more than likely that the baby gorilla would turn out to be a baby chimpanzee. However, I could not afford to turn the offer down, in case it really *was* a baby gorilla.

'How much are you asking for it?' I inquired, and took a firm grip of the telephone.

'Twelve hundred pounds,' said the dealer.

A brief vision of my Bank Manager's face floated before my eyes, and I repressed it sternly.

'All right,' I said, in what I hoped was a confident voice, 'I'll meet it at London Airport and, if it's in good condition, I'll have it.'

I put down the telephone to find Jacquie regarding me with a basilisk eye.

'What are you going to have?' she inquired.

'A baby gorilla,' I said nonchalantly.

'Oh, how lovely,' said Mother enthusiastically, 'they're such dear little things.'

Jacquie was more practical.

'How much?' she asked.

'As a matter of fact it's very reasonable,' I said, 'you know how rare gorillas are, and you know that our policy now is to concentrate on the rare things. I feel this is a wonderful opportunity . . .'

'How much?' Jacquie interrupted brutally.

'Twelve hundred pounds,' I replied and waited for the storm.

'Twelve hundred pounds? *Twelve hundred pounds?* You must be *mad*. You've got an overdraft the size of the National Debt and you go and say you'll pay twelve hundred pounds for a gorilla? You must be out of your mind. Where d'you think we're going to find twelve hundred pounds, for Heaven's sake. And what d'you think the Bank Manager's going to say when he hears? You must be stark staring *mad*.'

'I shall get the money from other sources,' I said austerely. 'Don't you realize that this island is infested with rich people who do nothing all day long but revolve from one cocktail party to another, like a set of Japanese waltzing mice. It's about time they made a contribution towards animal conservation. I shall ask them to contribute the money.'

'That's an even stupider idea than saying you'll have the gorilla in the first place,' said Jacquie.

Ignoring my wife's pessimistic and anti-social outlook, I picked up the telephone and asked for a number.

'Hallo? Hope? Gerry here.'

'Hallo,' said Hope resignedly. 'What can I do for you?'

'Hope, I want you to give me a list of all the richest people on the Island.'

'All the richest people?' said Hope in bewilderment, 'now *what* are you up to?'

'Well, I've just been offered a baby gorilla at a very reasonable price . . . twelve hundred pounds . . . only I don't happen to have twelve hundred pounds at the moment. . . .'

The rest of my sentence was drowned by Hope's rich laughter.

'So you hope to get the wealthy of the Island to buy it for you?' she said, chortling. 'Gerry, really, you're *dotty*.'

'I don't see what's wrong with the idea,' I protested, 'they should be glad to contribute towards buying such a rare creature. After all, if breeding colonies of things like gorillas aren't established in captivity soon, there won't be any left at all. Surely these people *realize* this?'

'I'm afraid they don't,' said Hope. 'I realize it and you realize it, but I'm afraid the average person either doesn't or couldn't care less.'

'I suppose you're right,' I said gloomily, 'anyway, I think it's worth a try, don't you?'

'It's worth a try, but I wouldn't pin too much faith to their generosity, if I were you,' said Hope. 'Anyway, give me half an hour and I'll ring you back.'

Half an hour later Hope dictated a list of about fifty people over the telephone, while I wrote them down feverishly. Then I looked up the telephone numbers, took a deep breath and started.

'Good morning, Mrs Macgurgle? Gerald Durrell from the Zoo here, I'm so sorry to worry you, but we've just been offered a baby gorilla . . . at a very reasonable price . . . twelve hundred pounds . . . well, yes, but it's not expensive for a gorilla . . . well, I was wondering if you'd care to purchase a small portion of it . . . say a leg or something? You would? That's immensely kind of you . . . thank you very much indeed . . . Good-bye.'

By lunchtime I had collected two hundred pounds. Only another thousand to go and the gorilla was mine. It was at this point that I discovered the next person on the list was Major Domo. I had never met him and I had no idea how he would react to the suggestion that he might buy a bit of gorilla. To my immense relief, the suggestion seemed to amuse him, for he chuckled.

'How much is it?' he asked.

'Twelve hundred pounds,' I said.

'And how much have you collected already?'

'Two hundred pounds.'

'Well,' said Major Domo, 'you'd better come along this afternoon and I'll find you the balance.'

To say I was speechless means nothing. When I had gone to the telephone I thought there might be a chance of getting twenty-five pounds, possibly even fifty. A hundred would have been beyond the dreams of avarice. And here was Major

Domo virtually handing me a baby gorilla on a platter, so to speak. I stammered my thanks, slammed down the telephone and rushed round the Zoo, telling everyone of the fact that we were going to have a baby gorilla.

The great day came and I flew over to London airport to collect the ape. My one fear now was that when I arrived there it would turn out to be a chimpanzee after all. The dealer met me and escorted me to a room in the R.S.P.C.A. Animal Shelter. He threw open the door, and the first thing I saw was a couple of baby chimpanzees sitting on a table meditatively chewing bananas. My heart sank, and I had visions of having to go back to Jersey empty-handed. But the dealer walked over to a crate in the corner, opened the door and N'pongo walked into my life.

He stood about eighteen inches high and was quite the most handsome and healthy-looking baby gorilla I had ever seen. He strolled stockily across the room towards me and then held up his arms to be lifted up. I was amazed at how heavy he was for his size, and I soon realized that this was all solid bone and muscle; there was not a spare ounce of fat on him. His light chocolate-coloured fur was thick and soft, and the skin on his hands, feet and face was soft and glossy as patent leather. His eyes were small and deep set, twinkling like chips of coal. He lay back in my arms and studied me carefully with an unwinking stare, and then lifted a fat and gentle forefinger and investigated my beard. I tickled his ribs and he wriggled about in my arms, giggling hoarsely, his eyes shining with amusement. I sat him down on a convenient table and handed him a banana which he accepted with little bear-like growlings of pleasure, and ate very daintily compared to the chimpanzees who were stuffing their mouths as full as they could. I wrote out the cheque and then we bundled N'pongo – growling protests – back into the crate, and went off to catch the plane for Jersey.

When we landed at the airport I took N'Pongo out of his crate and we drove to the Zoo with him sitting on my lap, taking a great interest in the cows we passed, and occasionally turning round so that he could peer up into my face. When we arrived I carried him up to our flat, for his cage was not quite ready and I had decided that he would have to spend a couple of days in our guest-room. His grave, courteous manner and his rather sad expression immediately won over both Jacquie and my mother, and before long he was lolling back on the sofa while they plied him with delicacies, and the staff came upstairs one by one to pay homage to him as though he was some black potentate. Having previously suffered by keeping Chumley the chimpanzee in the house, I knew from bitter experience that there was nothing like an ape for turning a civilized room into something closely resembling a bomb site in an incredibly short space of time, so I watched N'Pongo like a hawk. When he became bored with lying on the sofa, he decided to make a circuit of the room to examine anything of interest. So he walked slowly round like a small black professor in a museum, pausing now to look at a picture, now to stroke an ornament, but doing it so gently that there was never any danger that he would break anything. After the attitude I was used to with Chumley, I was captivated by N'Pongo's beautiful behaviour. You would have thought that he had been brought up in a house, to watch him. Apart from a slight lapse when he wet the floor (and he could not be expected to know that this was not done in the best circles), his behaviour was exemplary, so much so that, by the time we put him to bed, my mother was doing her best to try to persuade me to keep him in the house permanently. I had, however, learnt my lesson with Chumley, and I turned a deaf ear to her pleas.

N'Pongo, of course, did not leave the guest-room in the condition that he found it, but this was only to be expected.

Although his manners were exemplary, he was only a baby and could not be expected to automatically assume civilized behaviour, because he was living in the house. So the guest-room, when he left it, bore numerous traces of his presence: on one wall, for instance, was something that resembled a map of Japan drawn by one of the more inebriated ancient mariners. This was nicely executed in scarlet and was due to the fact that I had thought N'Pongo might like some tinned raspberries. He had liked them, and his enthusiasm at this new addition to his diet had resulted in the map of Japan. There was also straw. Next to paraffin, I know of no other commodity that manages to worm its way – in a positively parasitic fashion – into every nook and cranny before you are aware of it. For months after N'Pongo's sojourn in our guest-room we were apologizing to guests for the floor which, in spite of hoovering, looked like a sixteenth-century ale-house.

There was also the fact that the handle on the door drooped at rather a depressed angle since N'Pongo, after receiving his meal, had attempted to follow me out of the room. Knowing that the handle by some magical means opened and closed the door, but not knowing exactly how to manipulate it he had merely pulled it downwards with all his strength. As I tried, unsuccessfully, to bend the handle back into position again, I reflected that N'Pongo was only about two years old and that his strength would increase in proportion to his size.

One of the things which particularly interested me about him was his different approach to a problem or a situation. If, for example, a baby chimpanzee is used to being brought out of its cage, on being reincarcerated it will carry on like one of the more loquacious heroines in a Greek tragedy, tearing its hair, rolling with rage on the floor, screaming at the top of its voice and drumming its heels on every available bit of woodwork. N'Pongo was quite otherwise. Although deploring it, he would accept the necessity of being locked up again in his cage. He would try his best to divert you from this course of action, but when he realized that it had become inevitable he would submit with good grace. His only protest would be a couple of sharp and faintly peevish screams, as he saw you disappear, whereas the average chimpanzee of his age and with his background would have gone on having hysterics for a considerable length of time. Owing to his attractive appearance and disposition, his good manners and his very well-developed sense of humour, N'Pongo was in a very short space of time the darling of the Zoo. Every fine afternoon he was brought out on to the lawn in front of the yew hedge and there he would show off to his admirers: either lolling in the grass, with a bored expression on his face, or else, with a wicked gleam in his eye, working out how he could pose for his photograph to be taken by some earnest visitor and then rush forward at the crucial moment, grasp the unfortunate person's leg and push it from under him – a task that gave him immense amusement and generally resulted in the visitor sustaining a slipped disc and having an excellent picture of a completely empty section of lawn.

Within twelve months N'Pongo had almost doubled his size, and I felt it was now time to try, by fair means or foul, to obtain a mate for him. Unless they lack finances, I have no use for zoos that acquire animals purely for exhibition and make no effort to provide them with a mate; this applies particularly to apes. The problem does not arise while they are young, for they accept the human beings around them as their adopted, if slightly eccentric, family. Then comes the time when they are so powerful that you do not, if you have any intelligence, treat them in the same intimate way. When a gorilla or chimpanzee or an orang-utan at the age of three or four pulls your legs from under you, or jumps from a considerable height on to the back of your neck, it tests your stamina to the full and is done because you are the only companion with whom he has to play. If he is left on his own, and he is a nice-natured ape, he will try to play the same games with you when he is eleven and twelve; this means a broken leg or a broken neck. So if this friendly, exuberant animal is kept on his own and deprived of both the company of his own kind and that of human beings, you can hardly be surprised if he turns into a morose and melancholy creature. Not wanting to see N'Pongo degenerate into one of those magnificent but sad and lonely anthropoids that I have seen in many zoos (including some that had ample resources at their disposal for providing a mate), I thought the time had come to try to procure a wife for N'Pongo, even though I knew that our funds would probably not stretch that far. I telephoned the dealer from whom we had N'Pongo and asked him about the possibilities of obtaining a female gorilla. He told me he had just been offered one, a year or so younger than N'Pongo, but now, owing to the political state in Africa, the price had increased and he was asking £1,500. There followed two days of soul searching. I knew we could not afford that amount of money in a lump sum, but we might be able to do it if it were spread over a period. Once again I telephoned the dealer and asked him whether he would consider letting us have the animal on hire purchase terms. To his credit and to my relief he agreed and said that his representative would bring her over to Jersey in a week's time. The whole Zoo waited for that day with bated breath. I, prompted by a slightly acrimonious conversation with my bank manager, spent the week by having a collecting box made, over which hung the notice: 'We have bought Nandy on the instalment plan, please help us to keep up the instalments.'

So Nandy arrived, crouched in a crate that I would have considered small for a squirrel. She, like N'Pongo, appeared to be in perfect condition: her fur was glossy, she was fat, and her skin had a sheen like satin, but at first sight of her it was her eyes that impressed me most. N'Pongo's eyes, as I have said, were small and deep-set, calculating and full of humour. Nandy's eyes were large and lustrous, and when she looked sideways, she showed the whites of them; but they were frightened eyes that did not look at you squarely. They were the eyes of an animal that had had little experience of human beings, but even that limited experience had given her no reason to trust or respect them. When we released her from her cage, I could see the reason: right across the top of her skull was a scar which must have measured six or seven inches in length. Obviously, when she was being caught, some over-enthusiastic and intrepid human being had given her a blow with a machete which had split her scalp like a razor slash. It must have been a glancing blow, or else her skull would have been split in two. With such an introduction to the human race, you could hardly blame Nandy for being a little anti-social. This great slash was by now completely healed and there was only the long white scar to be seen through the hair of her head, which reminded me of the curious imitative and quite unnecessary partings that so many Africans carve in their hair with the aid of a razor.

We kept Nandy in a separate cage for twenty-four hours, so that she could settle down. The cage was next to N'Pongo's, to enable her to see her future husband, but she evinced as little interest in him as she evinced in us. If you tried to talk to her and looked directly at her face, her eyes would slide from side to side, only meeting yours for a sufficient length of time to judge what your next action might be. Eventually deciding that the wire between us rendered us comparatively harmless, and rather than look at us, she turned her back. She had such a woebegone, frightened face that one longed to be able to pick her up and comfort her, but she had been too deeply hurt, and this was the last thing she would have appreciated. It would take us, I reckoned roughly, at least six months to gain her confidence, even with the example of the pro-human N'Pongo.

The morning when we let her into the cage with N'Pongo was a red-letter day, but fraught with anxiety. He had by now become so well established and was such a fearfully extroverted character that he obviously considered he was the only gorilla in the world and that all human beings were his friends. We did not know how he would react to Nandy's sullenness and anti-human attitude. Although for twenty-four hours he could see her in the cage alongside him, he had shown absolutely no interest in her presence. Thus when the great moment of introduction came, we stood by with buckets of water, brushes, nets and long sticks, just in case the engagement party did not come off with the same romantic swing that one expects from reading women's magazines. When all was ready, we opened the shutters and Nandy, looking thoroughly distrustful, sidled her way from the small cage into N'Pongo's comparatively palatial quarters. There she put her back to the wall and squatted, her eyes darting to and fro with a curiously suspicious, yet belligerent, expression on her face. Now she was actually in the cage with N'Pongo, who was sitting up on a branch, watching her with the same expression of uninterested mistrust that he reserved for some new item of diet, we could see that she was very much smaller than he – in fact only about half his size. They sat steadily contemplating each other for some minutes, while we on the other side of the wire did hasty checks to make sure that all our buckets of water, nets, sticks and so on were easily accessible.

This was the critical moment: the two gorillas and ourselves were frozen into immobility. To any spectator, who did not know the circumstances, we would have appeared like one of the more bizarre of Madame Tussaud's exhibits. Then N'Pongo stretched out a black hand with fingers the size of chipolata sausages, clasped the wire and rolled himself carefully to the ground. There he paused and examined a handful of sawdust, as though it were the first time he had ever come across such a commodity. Then, in a casual, swaggering manner, he sauntered in a semi-circle which took him close to Nandy, and then, without looking at her, but with the utmost speed, he reached out a long, powerful arm, gripped a handful of her hair and pulled it, and then hurried along the perimeter of the cage as though nothing had happened. Nandy by nature has always been – and I fear will always be – a little slow in the uptake, and so N'Pongo was some six feet away before she realized what had happened. By then the baring of her teeth and her grunt of indignation were quite useless. The first round, therefore, went to N'Pongo, but before he could get exalted views of male superiority I felt that we should bring up our second line of defence. We removed the buckets and nets and produced two large dishes full of a succulent selection of fruits. One was presented to N'Pongo and one to Nandy.

There was one slight moment of tension when N'Pongo having examined his own plate, decided that possibly Nandy' contained additional delicacies, which his lacked, and went of to investigate. Nandy, however, was still smarting under th indignity of having had her hair pulled and she greete N'Pongo's investigation of her plate with such a display o belligerence that N'Pongo, being essentially a good humoure and cowardly creature, retreated to his own pile of food For the next half an hour they both fed contentedly at op posite ends of the cage. That night N'Pongo, as usual, slept o his wooden shelf, while Nandy, looking like a thwarted suffra gette, curled up on the floor. All through the following da they had little jousts with each other to see who would occup what position. They were working out their own protocol should Nandy be allowed to swing on the rope when N'Pong was sitting on the cross beam? Should N'Pongo be allowed t pinch Nandy's carrots even though they were smaller than hi own? It had the childishness of a General Election but wa three times as interesting. However, by that evening, Nand had achieved what amounted to Votes for Female Gorillas and both she and N'Pongo shared the wooden shelf. Judgin by the way N'Pongo snuggled up to her, he was not at a averse to this invasion of his bedroom.

It was obvious from the first that the marriage of th gorillas was going to be a success. Although they were s different in character, they quite plainly adored one another N'Pongo was the great giggling clown of the pair, whil Nandy was much quieter, more introspective and watchful N'Pongo's bullying and teasing of her was all done withou any malice and out of a pure sense of fun, and this Nand seemed to realize. Occasionally, however, his good humoure teasing would drive her to distraction: it must have bee rather like being married to a professional practical joker When she reached the limit of her endurance, she would los her temper, and with flashing eyes and open mouth would chase him round the cage while he ran before her, giggling hysterically. If she caught him, she would belabour him with her fists while N'Pongo lay on the ground curled up in a ball. Nandy might just as well have tried to hurt a lump of cement – in spite of her strength – for he would just lie there, laughing to himself, his eyes shining with good humour. As soon as she tired of trying to make some impression on his muscular body, Nandy would stalk off to the other end of the cage, and N'Pongo would sit up, brush the sawdust from himself, beat a rapid tattoo of triumph on his breast or stomach and then sit there with his arms folded, his eyes glittering, working out what other trick he could play to annoy his wife.

To have acquired such a pair of rare and valuable animals was, I considered, something of an achievement, but now, I discovered, we were to live in a constant state of anxiety over their health and well-being: every time one of them got sawdust up its nose and sneezed, we viewed this with alarm and despondency – was it a prelude to pneumonia or something worse? The functioning of their bowels became a daily topic of conversation. I had had installed a magnificent communication apparatus in the Zoo, for, small though it was, it could take a considerable length of time to locate the person required at the moment you wanted him. So at various salient points throughout the grounds, small black boxes were screwed to the walls, through which the staff could speak with the main office and vice versa. One of these boxes was also installed in our flat, so that I could be apprised of what was going on and be warned should any crisis arise. The occasion when I had doubts as to the wisdom of this system was the day when we were entertaining some people that we did not know very well. In the middle of one of those erudite and futile conversations one has to indulge in, the black box on the bookcase

gave a warning crackle, and, before I could leap up to switch
it off, a sepulchral and disembodied voice said:

'Mr Durrell, the gorillas have got diarrhoea again.'

I know of no equal to this remark for putting a blight on a
party. However, N'Pongo and Nandy grew apace, and to our
intense relief developed none of the diseases that we feared
they might contract.

Then came N'Pongo's first real illness. I had just arranged
to spend three weeks in the South of France, which was to be
a sort of working holiday, for we were to be accompanied by
a B.B.C. producer whom I hoped to convince of the necessity
for making a film about life in the Camargue. Hotels had been
booked, numbers of people, ranging from bull-fighters to
ornithologists, had been alerted for our coming, and everything seemed to be running smoothly. Then four days before
we were due to depart, N'Pongo started to look off colour.
Gone was his giggling exuberance; he lay on the floor or on
the shelf, his arms wrapped round himself, staring into space,
and only just taking enough food and milk to keep himself
alive. The only symptom was acute diarrhoea. Tests were
hurriedly made and the advice of both the vet and medical
profession acted upon, but what he was suffering from remained a mystery. As with all apes, he lost weight with horrifying rapidity. On the second day he stopped eating altogether and even refused to drink his milk, so this meant we
could administer no antibiotic. Almost as you watched, his
face seemed to shrink and shrivel and his powerful body grow
gaunt. What had once been a proudly rotund paunch now
became a ghastly declivity where his ribs forked. Now his
diarrhoea was quite heavily tinged with blood and at this
symptom I think most of us gave up hope. We felt that if he
would only eat something, it might at least give him some
stamina to withstand whatever disease he was suffering from,
as well as rouse him out of the terrible melancholia into which

he was slipping, as most of the anthropoids do when they are ill.

Jacquie and I went down to the market in St Helier and there we walked among the multi-coloured stalls that surround the charming, Victorian fountain with its plaster cherubim, its palms and its maidenhair fern and its household cavalry – the plump scarlet goldfish. It was difficult to know what to choose for N'Pongo that would tempt his appetite, for he had such an excellent variety of food in his normal diet. So we bought out-of-season delicacies that cost us a small fortune. Then, when we were loaded down with exotic fruits and vegetables, I suddenly noticed on a stall that we were passing an immense green and white water melon. Water melon is not to everyone's taste, but I personally prefer it to ordinary melon. It occurred to me that the bright pink-coloured, scrunchy, watery interior with its glossy black seeds might be something that would appeal to N'Pongo, for, as far as I knew, he had never sampled it before. We added the gigantic melon to our loads and drove back to the Zoo.

By now, through lack of food and drink, N'Pongo was in a very bad way. Jeremy had managed to persuade him to drink a little skimmed milk by the subterfuge of rubbing a Disprin on his gums. The Disprin, of course, dissolved rapidly and the taste not being to his liking, N'Pongo was only too happy to take a couple of gulps of the milk to wash out his mouth. One by one we presented him with the things we had obtained in the market, and one by one he viewed them with an apathetic glance and refused the hothouse grapes, the avocado pears and other delicacies. Then we cut him a slice of water melon, and for the first time he displayed signs of interest. He prodded the slice with his finger and then leant forward and smelt it carefully. The next minute he had the slice in his hands, and to our great delight had started to eat. But we did not become too jubilant, for we knew that the water melon contained practically no nutriment, but at least it had aroused his interest in food again. The next thing was to try to administer an antibiotic, as by now the consensus of expert opinion was that he was suffering from a form of colitis. Since he still refused to take any quantity of liquid in which we could mix medicines, there was only one way to get the antibiotic into him, and that was by injection.

We enticed N'Pongo out of his cage and kept Nandy shut up; he would be sufficiently difficult to deal with, in spite of his emaciated condition, without having any assistance from his, by now, extremely powerful wife. He squatted on the floor of the Mammal House, staring about him with dull, sunken eyes. Jeremy squatted one side of him, with a supply of water melon to try to maintain his interest, while I on the other side hastily prepared the syringe for the injection. N'Pongo watched my preparations with a mild interest and once put out his hand gently to try to touch the syringe. When I was ready, Jeremy endeavoured to distract his attention with pieces of water melon, and as soon as his head was turned away from me I pushed the needle into his thigh and pressed the plunger home. N'Pongo gave no sign of having even noticed this. He followed us obediently back into his cage and, with a small piece of water melon, retired to his shelf where he curled up on his side, his arms folded, and stared at the wall. The following morning he showed very slight signs of improvement, and using the same subterfuge we managed to give him another injection. For the rest of the day there seemed no change in him, and although he ate some of the melon and drank a little skimmed milk he did not show any radical signs of progress.

I was now in a quandary: in twenty-four hours' time I was due to leave for France. There I had organized and stirred up a bees' nest of helpers and advisers. The B.B.C. were also under the impression that the trip was a foregone conclusion. If I put it off at this juncture, I would have put a tremendous amount of people to a lot of trouble for nothing, and yet I felt I could not leave N'Pongo unless I was satisfied that he was either on the mend or else beyond salvation.

Then, the day before I was due to leave, he suddenly turned the corner. He started drinking his Complan – a highly concentrated form of dried milk – and eating a variety of fruits. By the evening of that day he showed considerable signs of improvement and had eaten quite a bit of food. The next morning I went down very early to look at him, for I was due to catch my plane to Dinard at eight-thirty. He was sitting up on the shelf, and although he still looked emaciated and unwell his eyes had a sparkle in them that had been lacking for the past few days. He ate quite well and drank his Complan, and I felt that he was at last on the road to recovery. I drove down to the airport and caught the plane to Dinard, and we motored down to the South of France. It cost a small fortune in trunk-calls to Jersey to keep myself apprised of N'Pongo's progress, but, every time I telephoned, the reports got better and better, and when Jeremy informed me that N'Pongo had drunk one pint of Complan and eaten three slices of water melon, two bananas, one apricot, three apples and the white of eight eggs, I knew there was no further cause for alarm.

By the time I returned from France, N'Pongo had put on all the weight he had lost, and when I went into the Mammal House there he was to greet me, his old self: massive, black and rotund, his eyes glittering mischievously as he tried to inveigle me close enough to the wire so that he would pull the buttons off my coat. I reflected, as I watched him rolling on his back and clapping his hands vigorously in an effort to attract my attention, that, though it was delightful to have creatures like this – and of vital importance that they should be kept and bred in captivity – it was a two-edged sword, for the anxiety you suffered when they became ill made you wonder why you started the whole thing in the first place.

CHAPTER EIGHT

ANIMALS IN TRUST

Dear Mr Durrell,
You will probably be astonished to receive a letter from a complete
stranger . . .

THE Zoo has now been in existence for five years. During that time we have worked steadily towards our aim of building up our collection of those animals which are threatened with extinction in the wild state. Examples of these are our chimpanzees, a pair of South American tapirs, but, perhaps, the pair of gorillas are one of the most important of our acquisitions and one of which we are extremely proud. Apart from these, we have over the past year obtained a number of valuable creatures. It is not always possible to buy or collect these animals, so recently in exchange for an ostrich we had our binturong, a strange, small, bear-like animal with a long prehensile tail, which comes from the Far East; and a spectacled bear, whom we have christened Pedro.

Spectacled bears are the only members of the family to be found in South America, inhabiting a fairly restricted range high in the Andes. They are a blackish brown colour with fawn or cinnamon spectacled markings round the eyes and short waistcoats of a similar colour. They grow to be as large as the ordinary black bear, but Pedro, when he arrived, was still quite a baby and only about the size of a large retriever. We soon found that he was ridiculously tame and liked nothing better than to have his paws held through the bars while he munched chocolate in vast quantities. He is an incredible pansy in many ways, and several of the attitudes he adopts – one foot on a log while he leans languidly against the bars of his cage, with his front paws dangling limply – remind one irresistibly of the more vapid and elegant young men one can see at cocktail parties. He very soon discovered that if he did certain tricks the flow of chocolates and sweets increased a hundredfold, and so he taught himself to do a little dance. This consisted of standing on his hind legs and bending over backwards as far as he could, without actually falling, and then revolving slowly – a sort of backward waltz. This never failed to enchant his audience. To give him something with which to amuse himself, we hung a large empty barrel from the ceiling of his cage, having knocked both ends out of it: this formed a sort of circular swing and gave Pedro a lot of pleasure. He would gallop round the barrel and then dive head first into it, so that it swung to and fro vigorously. Occasionally, he would dive a bit too strenuously and come shooting out of the other end of the barrel and land on the ground. At other times, when he was feeling in a more soulful mood, he would climb into his barrel and just lie there, sucking his paws and humming to himself, an astonishingly loud vibrant hum as though the barrel contained quite a large dynamo.

Pedro was, at first, in temporary quarters, but, as we hoped to get him a mate eventually, we had to build him a new cage. During the period while his old quarters were being demolished and his new one being erected, he was confined in a large crate to which, at first, he took grave exception. However, when we moved it next to one of the animal kitchens and the fruit store, he decided that life was not so bad after all. The staff were constantly in view and nobody passed his crate without pushing a titbit to him through the bars. Then, two days before he was due to be moved into his new home, it happened. Jacquie and I were up in the flat, having a quiet cup of tea with a friend, when the inter-communication crackled and Catha's voice, as imperturbable as though she were announcing the arrival of the postman, said:

'Mr Durrell, I thought you would like to know Pedro is out.'

Now, although Pedro had been small when he arrived, he had grown with surprising rapidity and was now quite a large animal. Also, although he still appeared ridiculously tame, bears, I am afraid, are some of the few creatures in this world which you cannot trust in any circumstance. So, to say that I was alarmed by this news would be putting it mildly. I fled downstairs and out of the back door. Here, where the animal kitchen and fruit store form an annexe with a flat roof, I saw

Pedro. He was galloping up and down on the roof, obviously having the time of his life. The unfortunate thing was that one of the main windows of the flat overlooked this roof, and if he went through that he could cause a considerable amount of havoc in our living quarters. Pedro was plainly unfamiliar with the substance called glass, and, as I watched, he bounded up to the window, reared up on his hind legs and hurled himself hopefully forward. It was lucky that it was an old-fashioned sash window with small panes of glass, and this withstood his onslaught. If it had been one big sheet of glass, he would have gone straight through it and probably cut himself badly. But with a slightly astonished expression on his face he rebounded from it: what appeared to be a perfectly good means of getting into the flat was barred by some invisible substance. I rushed round to where the crate was, in an endeavour to lift up the sliding door which, as always happens in moments of this sort of crisis, stuck fast. Pedro came and peered at me over the edge of the roof and obviously thought that he should come down to my assistance, but the long drop made him hesitate. I was still struggling with the door of the cage when Shep appeared with a ladder.

'We'll never get him down without this,' he said, 'he's frightened to jump.'

He placed the ladder against the wall, while I continued my struggles with the door of the crate. Then Stefan came on the scene and was coming to my aid, when Pedro suddenly discovered the ladder. With a little 'whoop' of joy, he slid down it like a circus acrobat and landed in an untidy heap at Stefan's feet.

Now, Stefan was completely unarmed and so was I, but fortunately Stefan kept his head and did the right thing: he stood absolutely still. Pedro righted himself and, seeing Stefan standing next to him, gave a little grunt, reared up on his hind legs and placed his paws on Stefan's shoulders, who went several shades whiter but still did not move. I looked round desperately for some sort of weapon with which I could hit Pedro, should this be a preliminary to an attack on Stefan. Pedro, however, was not interested in attacking anyone. He gave Stefan a prolonged and very moist kiss with his pink tongue and then dropped to all fours again and started galloping round and round the crate, like an excited dog. I was still trying ineffectually to raise the slide when Pedro made a miscalculation. In executing a particularly complicated and beautiful gambol, he rushed into the animal kitchen. It was the work of a second for Shep to slam the door, and we had our escapee safely incarcerated. Then we freed the reluctant slide and pushed the crate up to the kitchen, opened its door, and Pedro re-entered his quarters without any demur at all. Stefan vanished and had a strong cup of tea to revive himself. Two days later we released Pedro into his spacious new quarters, and it was a delight to watch him rushing about, investigating every corner of the new place, hanging from the bars, pirouetting in an excess of delight at finding himself in such a large area.

When owning a zoo, the question of Christmas, birthday and anniversary presents is miraculously solved: you simply give animals to each other. To any harassed husband who has spent long sleepless nights wondering what gift to present to his wife on any of these occasions, I can strongly recommend the acquisition of a zoo, for then all problems are answered. So, having been reminded by my mother, my secretary and three members of the staff that my twelfth wedding anniversary was looming dark and forbidding on the horizon, I sat down with a pile of dealers' lists, to see what possible specimens I could procure that would have a two-fold value of both gladdening Jacquie's heart and enhancing the Zoo. The whole subterfuge had this additional advantage: I could spend far more money than I would have done otherwise, without the risk of being nagged for my gross extravagance. So, after several mouth-watering hours with the lists, I eventually settled on two pairs of crowned pigeons, birds which I knew Jacquie had always longed to possess. They are the biggest of the pigeon family and certainly one of the most handsome, with their powder-blue plumage, and scarlet eyes and their great feathery crests. Nobody knows how they are faring in the wild state, but they seem to be shot pretty indiscriminately both for food and for their feathering, and it is quite possible, before many years have passed, that crowned pigeons will be on the danger list. I saw that at that precise moment the cheapest crowned pigeons on the market were being offered by a Dutch dealer. Fortunately, I have a great liking for Holland and its inhabitants, so I thought it would be as well if I went over personally to select the birds, for, as I argued to myself, it would enable me to choose the very finest specimens (and for a wedding anniversary, surely nothing but the best would do?), and at the same time give me a chance to visit some of the Dutch zoos which are, in my opinion, among the finest in the world. Having thus salved my conscience, I went across to Holland.

It was just unfortunate that the very morning I called at the dealers to choose the crowned pigeons, a consignment of orang-utans had arrived. This put me in an extremely awkward position. First, I have always wanted to have an orangutan. Secondly, I knew that we could not possibly afford them. Thirdly, owing in part to this trade in these delicate and lovely apes, their numbers have been so decimated in the wild state that it is possible within the next ten years they may well become extinct. As an ardent conservationist what was I to do? I could not report the dealer to anyone, for the simple reason that now they had managed to reach Holland there was no law against him having them in his possession. I was in a quandary. I could either not even look at the apes and leave them to his tender mercies, or else I could, as it were obliquely, encourage a trade of which I strongly disapproved, by rescuing them.

By this time I was so worked up over the conservation aspect of this problem that the financial side of it had disappeared completely from my mind. Knowing full well what would happen, I went and peered into the crate containing the baby orang-utans and was immediately lost. They were both bald and oriental-eyed; the male, who was the slightly larger of the two, looked like a particularly malevolent Mongolian brigand, while the female had a sweet and rather pathetic little face. As usual, they had great pot-bellies, owing to the ridiculous diet of rice on which the hunters and dealers insist on feeding them and which does them no good whatsoever, except to distend their stomachs and give them internal disorders.

They crouched in the straw, locked in each others arms; to each the other was the one recognizable and understandable

thing in a horrifying and frightening world. They both looked healthy, apart from their distended tummies, but they were so young I knew the chances of their survival were risky. The sight of them, however, clutching each other and staring at me with such obvious terror, decided me, and (knowing that I should never hear the end of it) I sat down and wrote out a cheque.

That evening I telephoned the Zoo to tell Jacquie that all was well and that I had not only managed to buy the crowned pigeons she wanted, but also two pairs of very nice pheasants. On hearing this, both Catha and Jacquie said that I should not be allowed to go shopping in animal dealers by myself and I had no sense of economy and why was I buying pheasants when I knew the Zoo could not afford them, to which I replied that they were rare pheasants and that was sufficient excuse. I then carelessly mentioned that I had also bought something else.

What, they inquired suspiciously, had I bought?

'A pair of orang-utans,' I said airily.

'Orang-utans?' said Jacquie. 'You must be mad. How much did they cost? Where are we going to keep them? You must be out of your mind.'

Catha, on being told the news, agreed with her. I explained that the orang-utans were so tiny that they would practically fit in your pocket and that I could not possibly leave them to just die in a dealer's shop in Holland.

'You'll love them when you see them,' I said hopefully, to which Jacquie's answer was a derisive snort.

'Well,' she said philosophically, 'if you have bought them, you have bought them, and I suggest you come back as quickly as possible before you spend any more money.'

'I am returning tomorrow,' I replied.

So the following day I sent the crowned pigeons and the pheasants off by air and travelled myself by sea with my two waifs. They were very suspicious and timid, although the female was more inclined to be trusting than the male, but after a few hours of coaxing they did take titbits from my hand. I decided after much deliberation to call the male Oscar and the female Bali, since it had some vague connotation with the area of the world from which they originated. Little was I to know that this would give rise to Jeremy perpetrating a revolting pun 'that when Oscar was wild, this made Bali high'.

I had decided to travel by sea with them because, first of all, I never travel by air if I can possibly avoid it. I am absolutely convinced that every aeroplane pilot who flies me has just been released from Broadmoor, suffering from acute *angina pectoris*. Also I felt the trip would be more leisurely and would give me a chance to establish some sort of contact with my charges. As regards the latter, I was perfectly correct: Bali had begun to respond quite well and Oscar had bitten me twice by the time I arrived.

As I anticipated, as soon as I returned to the Zoo with my two bald-headed, pot-bellied, red-haired waifs, everyone immediately fell in love with them. They were crooned over and placed in a special cage which had been prepared for them in advance, and hardly a moment of the day passed without someone or other going to peer at them and give them some delicacy. It was a month before they showed signs of recovering their self-confidence and began to realize that we were not the ogres they thought. Then their personalities blossomed forth and they very soon became two of the most popular inmates of the Zoo. I think it was their bald heads, their strange slant eyes, and their Buddha-like figures that made them so hilariously funny to watch as they indulged in the most astonishing all-in-wrestling matches that I have ever seen. Owing to the fact that their hind legs can, it seems, swivel round and round on the ball and socket joint of the hip in a completely unanatomical manner, these wrestling matches had to be seen to be believed. Gasping and giving hoarse chuckles, they would roll over and over in the straw, banging their great pot-bellies together, and so inextricably entwine their arms and legs that you began to wonder how on earth they would ever disentangle themselves. Occasionally, if Oscar became too rough, Bali would protest: a very reedy, high-pitched squeak which was barely audible and quite ridiculous from an animal of that size.

They grew at an astonishing rate and very soon had to be moved into a new cage. Here Jeremy had designed and had had constructed for them a special piece of furniture for their edification. It was like a long iron ladder slung from the ceiling. This gave them masses of handholds and they enjoyed it thoroughly; they took so much exercise on this that their tummies soon reduced to a more normal size.

In character they were totally different. Oscar was a real toughy; he was a terrible coward, but never lost an opportunity for creating a bit of mischief if he could. He is definitely the more intelligent of the pair and has shown his inventive genius on more than one occasion. In their cage is a recessed window; the window ledge we had boarded over to form a platform on which they could sit, and, leading up to it, an iron-runged ladder. Oscar decided (for some reason best known to himself) that it would be a good idea to remove all the boards from the window-sill. He tried standing on them and tugging, but his weight defeated his object. After considerable thought, he worked out the following method of dislodging the planks, which is one of the most intelligent things I have seen done by an ape. He found out that the top rung of the iron ladder lay some two inches below the overlap of the shelf. If he could slide something into this gap and press it downwards it would act as a lever, using the top rung of the ladder as fulcrum; and what better tool for his purpose than his stainless-steel dish? By the time we had found out that we had a tool-using ape in our midst, Oscar had prized up six of the boards and was enjoying himself hugely.

It is unfortunate that, like many apes, Oscar and Bali have developed some rather revolting characteristics, one of which is the drinking of each other's urine. It sounds frightful, but they are such enchanting animals and do it in such a way that you can only feel amused to see Oscar sitting up on his iron ladder urinating copiously, while Bali sits below with open mouth to receive the nectar, and then savours it with all the air of a connoisseur. She puts her head on one side, rolling the liquid round her mouth as if trying to make up her mind from which vineyard it came and in what year it was bottled. They also, unfortunately, enjoy eating their own excreta. As far as I know, these habits in apes apply only to specimens in captivity. In the wild state apes are on the move all the time and to a greater or lesser extent are arboreal, so that their urine and faeces drop to the forest floor below, and therefore they are not tempted to test their edibility. Once they start this habit in captivity, it is virtually impossible to break them of it. It does not appear to do them any harm, except, of course, that if they do happen to be infected with any internal parasites (of which you are endeavouring to cure them), they are constantly reinfecting themselves and each other by these means.

Other new arrivals of great importance from the point of view of conservation were a pair of tuataras from New Zealand. These astonishing reptiles had at one time had a wide range but were exterminated on the mainland, and are now found only in a few scattered groups of small islands off the coast of New Zealand. They are rigidly protected by the New Zealand Government and, only occasionally, the odd specimen is exported for some zoo. On a brief visit I paid to that country, I explained to the authorities the work I was trying to do in Jersey and they – somewhat unwisely – asked me if there was any member of the New Zealand fauna which I would particularly like to have. Resisting the impulse to say 'everything' and thus appear greedy, I said that I was very interested in tuatara. The Minister concerned said that he was sure they could see their way to letting me have one, to which I replied that I was not interested in having one, although this seemed like looking a gift horse in the mouth. I explained that my idea was to build up breeding colonies, and it was difficult, to say the least, to form a breeding colony with one animal. Could I, perhaps, have a pair? After due deliberation, the authorities decided that they would let me have a true pair of tuatara. This was indeed a triumph, for, as far as I know, we are the only Zoo in the world to have been allowed to have a true pair of these rare reptiles.

The climate of New Zealand is not unlike that of Jersey. Previously, when I had seen tuataras at various zoos, they had always been incarcerated in Reptile Houses in cages, the temperature of which fluctuated between 75° and 80° F. At the time this had not occurred to me as being a bad thing, but when I went to New Zealand and saw the tuataras in their wild state, I suddenly realized that the mistake the majority of European zoos had been making was to keep the tuatara as though it were a tropical reptile: this accounted for the fact that very few of these creatures kept in Europe had lived for any great length or time. Having obtained permission to have a pair of tuatara, I was quite determined that their cage must be the best possible, and that I would keep them at temperatures as near to the ones to which they were accustomed as we could manage. So when I was alerted from the Wildlife Department of New Zealand that the tuataras would be sent to me very shortly, we started work on their housing. This, in fact, resembles a rather superior greenhouse: it is twenty-one feet in length and eleven feet wide, with a glass roof. This roof is divided into windows, so that we can keep a constant current of air flowing through the cage and thus make sure that the temperature does not rise too much. A large quantity of earth and rockwork was then arranged and planted out, so as to resemble as closely as possible the natural habitat of the reptiles. We sank one or two pipes into the earth to act as burrows, should the tuataras not feel disposed to make their own, and then we waited for their arrival excitedly.

At last the great day came and we went down to the airport to collect them. They were carefully packed in a wooden box, the air holes of which did not allow me to see if they had survived the journey, and I remained in a state of acute frustration all the way back from the airport to the Zoo. There I could lay my hand on a screwdriver and remove the lid of the box, to see how our new arrivals had fared on their travels. As we removed the last screw and I prepared myself to lift the lid off, I uttered up a brief prayer. I lifted off the lid, and there, gazing at me benignly from the depths of the container, were a pair of the most perfect tuatara I had ever beheld. In shape they resemble a lizard, though anatomically they are so different that they occupy a family all of their own. They have, in fact, come down from prehistoric times virtually unchanged, and so if anything in the world can be dignified with the term prehistoric monster, the tuatara can.

They have enormous, lustrous dark eyes and a rather pleasant expression. Along the back is a fringe of triangular spines, white and soft, rather like the frill on a Christmas cake. This is more accentuated in the male than in the female. A similar row of spines decorates the tail, but these are hard and sharp, like the spikes on the tail of a crocodile. Their bodies are a sort of pale beige, mottled with sage green and pale yellow. They are, altogether, very handsome creatures with an extremely aristocratic mien.

Before releasing the tuataras into their new home, I wanted

o be sure that the journey had not upset them too much, and nat they would feed, so we left them in their travelling box ver-night and put twelve dead baby rats in with them. The ext day, to my delight, the box contained no trace of baby ats but a couple of rather portly and smug tuataras. It was bvious that a plane journey of thousands of miles to creatures f such ancient lineage was a mere nothing, and so we released nem into their new quarters. Here, I am glad to say, they have ettled down very well and have now grown so tame that they ill feed from your hand. I hope that in the not too distant uture we may make zoological history by breeding them, for, s far as I know, no zoo outside Australia and New Zealand ad succeeded in hatching baby tuataras.

Now that the Zoo was solvent and had acquired so many airs of threatened species, I felt the time had come to take the ext big step forward. It was essential, if we were to do the ork of saving threatened species which was my aim, that we ad to have outside financial assistance and that the whole peration had to be put on an intelligent scientific footing.

The answer, therefore, was for the Zoo to cease being a limited company and to become a proper scientific Trust.

On the face of it, this seems a fairly simple manoeuvre, but in practice it is infinitely more difficult. First you have to gather together a council of altruistic and intelligent people who believe in the aims of the Trust, and then launch a public appeal for funds. I shall not go into all the wearisome details of this period, which can be of no interest to anyone but myself. Suffice it to say that I managed to assemble a council of hard working and sympathetic people on the Island, who did not consider my aims so fantastic as to qualify me for a lunatic asylum, and with their help the Jersey Wildlife Preservation Trust came into being. We launched a public appeal for funds, and once more the people of Jersey came to my rescue, as they had done in the past with calves, or tomatoes, or snails, or earwigs. This time they came forward with their cheque books, and before long the Trust had acquired sufficient money to take over the Zoo.

This means that after twenty-two years of endeavour I shall have achieved one of the things that I most desired in the world, and that is to help some of the animals that have given me so much pleasure and so much interest during my lifetime. I realize that the part we can play here is only a very small one, but if by our efforts we can prevent only a tiny proportion of threatened species from becoming extinct, and, by our efforts, interest more people in the urgent and necessary work of conservation, then our work will not have been in vain.

FINAL DEMAND

Dear Sir,
Once again may we point out that your account is still overdrawn . . .

DON'T know whether you, who are reading this, have read
my of my other books, but if you have, or if you have only
ead this one with pleasure, it is the animals that have made it
njoyable for you. Whether your work is in the countryside,
r whether it is in a broker's office, or in a factory, animals –
lthough you may not realize it – like the forests and fields in
his world, are of importance to you, if only for the reason that
hey provide people like myself with material to write about,
which entertains you. A world without birds, without forests,
without animals of every shape and size, would be one that I,
ersonally, would not care to live in and which, indeed, it
would be impossible for man to live in. The rate of man's
rogress and, in consequence, his rape of an incredibly
eautiful planet accelerates month by month, and year by year.
t is up to everyone to try to prevent the awful desecration
f the world we live in, which is now taking place, and every-
ody can help in this in however a humble capacity. I am
oing what I can in the only way that I know, and I would
ke your support. As a rule, I frown on touting, but in a case
s urgent and as necessary as this, I throw my scruples over-
oard. If you feel that you want to help in this work, please
write to the address overleaf:

MENAGERIE MANOR

Jersey Wildlife Preservation Trust
c/o Jersey Zoological Park
Les Augres Manor
Trinity
Jersey
Channel Islands.

In the meantime, while there are still animals and green places left in the world, I shall do my best to visit them and write about them.

MORE ABOUT PENGUINS
AND PELICANS

Penguinews, which appears every month, contains details of all the new books issued by Penguins as they are published. It is supplemented by our stocklist which includes around 5,000 titles.

A specimen copy of *Penguinews* will be sent to you free on request. Please write to Dept EP, Penguin Books Ltd, Harmondsworth, Middlesex, for your copy.

In the U.S.A.: For a complete list of books available from Penguins in the United States write to Dept CS, Penguin Books, 625 Madison Avenue, New York, New York 10022.

In Canada: For a complete list of books available from Penguins in Canada write to Penguin Books Canada Ltd, 2801 John Street, Markham, Ontario L3R 1B4.

Gerald Durrell

THE DRUNKEN FOREST

'Once again he has returned from his explorations with another fascinating account of his adventures. He has the ability of overcoming any ignorance of or indifference to his own subject, and he will soon have the most disinterested reader crowing with delight over the habits of orange armadillos, horned toads, Budgett's, and other endearing animals – once you get to know them.

'On this occasion it is into the Argentine pampas and the little-known Chaco territory of Paraguay where Mr Durrell and his wife go wandering off . . . His sympathy with the animal world encourages the Disney in every creature to show itself' – *Time and Tide*

also published in Penguins

THE BAFUT BEAGLES
ENCOUNTERS WITH ANIMALS
MY FAMILY AND OTHER ANIMALS
THREE SINGLES TO ADVENTURE
THE WHISPERING LAND
A ZOO IN MY LUGGAGE

JERSEY WILDLIFE PRESERVATION TRUST

Gerald Durrell writes:

Have you enjoyed this book?

If you have, it is the animals that have made this possible; and these animals are not just characters in a book: they *really* exist. But many of them will not exist for much longer unless they have your help.

All over the world the wildlife that I write about is in grave danger. It is being exterminated by what we call the progress of civilization. A great number of creatures will become extinct in a very short time if something is not done, and done swiftly.

Some time ago I created on the Island of Jersey a zoological park which is now the headquarters of the Jersey Wildlife Preservation Trust. Our aim is to create a sanctuary in which we can establish breeding colonies of these threatened species, so that, even if they become extinct in the wild state, they will not vanish forever.

To do this work money is required.

Therefore we need as many members as possible to join the Trust. It will cost you little, but you will be helping a cause that is of the utmost importance and urgency. I say 'urgency' advisedly, because, as you read this, yet another species is added to the danger list.

If animals I write about have given you pleasure, please join the Trust. The animals will be greatly indebted for every subscription received.

Full particulars can be obtained from:

> The Secretary
> Jersey Wildlife Preservation Trust
> Les Augres Manor
> JERSEY
> Channel Islands